Subhas Chandra Mukhopadhyay and Gourab Sen Gupta (Eds.)

Autonomous Robots and Agents

Studies in Computational Intelligence, Volume 76

Editor-in-chief
Prof. Janusz Kacprzyk
Systems Research Institute
Polish Academy of Sciences
ul. Newelska 6
01-447 Warsaw
Poland
E-mail: kacprzyk@ibspan.waw.pl

Further volumes of this series
can be found on our homepage:
springer.com

Vol. 57. Nadia Nedjah, Ajith Abraham and Luiza de Macedo Mourella (Eds.)
Computational Intelligence in Information Assurance and Security, 2007
ISBN 978-3-540-71077-6

Vol. 58. Jeng-Shyang Pan, Hsiang-Cheh Huang, Lakhmi C. Jain and Wai-Chi Fang (Eds.)
Intelligent Multimedia Data Hiding, 2007
ISBN 978-3-540-71168-1

Vol. 59. Andrzej P. Wierzbicki and Yoshiteru Nakamori (Eds.)
Creative Environments, 2007
ISBN 978-3-540-71466-8

Vol. 60. Vladimir G. Ivancevic and Tijana T. Ivacevic
Computational Mind: A Complex Dynamics Perspective, 2007
ISBN 978-3-540-71465-1

Vol. 61. Jacques Teller, John R. Lee and Catherine Roussey (Eds.)
Ontologies for Urban Development, 2007
ISBN 978-3-540-71975-5

Vol. 62. Lakhmi C. Jain, Raymond A. Tedman and Debra K. Tedman (Eds.)
Evolution of Teaching and Learning Paradigms in Intelligent Environment, 2007
ISBN 978-3-540-71973-1

Vol. 63. Wlodzislaw Duch and Jacek Mańdziuk (Eds.)
Challenges for Computational Intelligence, 2007
ISBN 978-3-540-71983-0

Vol. 64. Lorenzo Magnani and Ping Li (Eds.)
Model-Based Reasoning in Science, Technology, and Medicine, 2007
ISBN 978-3-540-71985-4

Vol. 65. S. Vaidya, L.C. Jain and H. Yoshida (Eds.)
Advanced Computational Intelligence Paradigms in Healthcare-2, 2007
ISBN 978-3-540-72374-5

Vol. 66. Lakhmi C. Jain, Vasile Palade and Dipti Srinivasan (Eds.)
Advances in Evolutionary Computing for System Design, 2007
ISBN 978-3-540-72376-9

Vol. 67. Vassilis G. Kaburlasos and Gerhard X. Ritter (Eds.)
Computational Intelligence Based on Lattice Theory, 2007
ISBN 978-3-540-72686-9

Vol. 68. Cipriano Galindo, Juan-Antonio Fernández-Madrigal and Javier Gonzalez
A Multi-Hierarchical Symbolic Model of the Environment for Improving Mobile Robot Operation, 2007
ISBN 978-3-540-72688-3

Vol. 69. Falko Dressler and Iacopo Carreras (Eds.)
Advances in Biologically Inspired Information Systems: Models, Methods, and Tools, 2007
ISBN 978-3-540-72692-0

Vol. 70. Javaan Singh Chahl, Lakhmi C. Jain, Akiko Mizutani and Mika Sato-Ilic (Eds.)
Innovations in Intelligent Machines-1, 2007
ISBN 978-3-540-72695-1

Vol. 71. Norio Baba, Lakhmi C. Jain and Hisashi Handa (Eds.)
Advanced Intelligent Paradigms in Computer Games, 2007
ISBN 978-3-540-72704-0

Vol. 72. Raymond S.T. Lee and Vincenzo Loia (Eds.)
Computation Intelligence for Agent-based Systems, 2007
ISBN 978-3-540-73175-7

Vol. 73. Petra Perner (Ed.)
Case-Based Reasoning on Images and Signals, 2008
ISBN 978-3-540-73178-8

Vol. 74. Robert Schaefer
Foundation of Global Genetic Optimization, 2007
ISBN 978-3-540-73191-7

Vol. 75. Crina Grosan, Ajith Abraham and Hisao Ishibuchi (Eds.)
Hybrid Evolutionary Algorithms, 2007
ISBN 978-3-540-73296-9

Vol. 76. Subhas Chandra Mukhopadhyay and Gourab Sen Gupta (Eds.)
Autonomous Robots and Agents, 2007
ISBN 978-3-540-73423-9

Subhas Chandra Mukhopadhyay
Gourab Sen Gupta
(Eds.)

Autonomous Robots and Agents

With 130 Figures and 22 Tables

Springer

Subhas Chandra Mukhopadhyay
Institute of Information Sciences
and Technology
Massey University (Turitea Campus)
Palmerston North
New Zealand
E-mail: S.C.Mukhopadhyay@massey.ac.nz

Gourab Sen Gupta
School of Electrical and Electronic
Engineering
Singapore Polytechnic
Singapore
E-mail: Sengupta@sp.edu.sg

ISBN 978-3-642-09249-7 e-ISBN 978-3-540-73424-6
ISSN print edition: 1860-949X
ISSN electronic edition: 1860-9503

Springer is a part of Springer Science+Business Media
springer.com

Softcover reprint of the hardcover 1st edition 2007

Cover design: deblik, Berlin

Guest Editorial

This special issue titled "Autonomous Robots and Agents" in the book series of "Studies in Computational Intelligence" contains the extended version of the papers selected from those presented at the 3rd International Conference on Autonomous Robots and Agents (ICARA 2006) which was held in December 11-14, 2006 at Massey University, Palmerston North, New Zealand. A total of 132 papers were presented at ICARA 2006, of which 30 papers have been selected for this special issue.

The first three papers are in the category of Search and Rescue. In the first paper D. A. Williamson and D. A. Carnegie have described the development of an embedded platform which is the core of an Urban Search and Rescue (USAR) system. An embedded computer, network camera, supervisory module and wireless gateway along with a distributed high efficiency power supply have resulted in a compact, inexpensive and efficient embedded platform for robotic USAR. In the next paper J. L. Baxter et al., have described a potential field based approach for multi-robot search and rescue operation. R. Jarvis and M. Marzouqi have described a search strategy based on a probabilistic model of target in which location and the partitioning of the environment are performed autonomously.

The next five papers are in the category of Localisation, Mapping and Navigation. J. Schmidt et al., have presented an approach for indoor mapping and localisation using sparse range data, acquired by a mobile robot equipped with sonar sensors. T. Taylor has applied high level understanding to visual localisation for mapping. He has outlined some simple geometric relationships that can be used to recover vertical edges from an image when the camera is tilted. By comparing the slopes of vertical edges in successive images, it is possible to calculate distances moved or angles of rotation. H. Yussof et al., have developed an optical three-axis tactile sensor for object handling tasks in humanoid robot system. The developed sensor is capable of clearly acquiring normal force and shearing force. Furthermore, it can also detect force at the *XYZ*-axes of a Cartesian coordinate. C. Astengo-Noguez and R. Brena-Pinero have presented a clustering method based on negotiation in flock traffic navigation and showed by simulation that a non-centred solution can be achieved letting individual vehicles to communicate and negotiate among them. M. Happlod and M. Ollis have presented a novel method for terrain classification for mobile robot navigation that eliminates the need for guessing which geometric features are relevant. The system learns to mimic the classification judgments of a human expert rather than applying explicit rules or thresholds to measured features.

S. M. R. Hasan and J. Potgieter have presented a design of an integrated circuit for robotic visual object position tracking. H. Andreasson et al., have proposed a non-iterative algorithm for 3D laser scans based on parameter-free vision-based interpolation method.

The next three papers are on mobile robots. G. Seet et al., have described the design and development of an interactive 3D robotic simulator to study behaviour of mobile robots in a cooperative environment. A.N. Chand and G.C. Onwubolo have developed a mobile robot intended to eliminate the human component in the book retrieval process by autonomously navigating to and retrieving a specific book from a bookshelf. In the next paper a trajectory planning strategy for surveillance mobile robots has been discussed by L.S. Martins-Filho and E.E.N. Macau.

The next group of papers are in the area of algorithms and simulation for robotic systems. T.Y. Li and C.C. Wang have proposed a genetic algorithm to generate an optimal set of weighting parameters for composing virtual forces within a given environment and the desired movement behaviour. J.A. Ferland et al., have applied the particle swarm algorithm in the capacitated open pit mining problem. N. Murata et al., have proposed a genetic algorithm based automatic adjustment method of the optical axis in laser systems. A novel genetic algorithm based method to evolve control system of a computer simulated mobile inverted pendulum has been presented by M. Beckerleg and J. Collins.

M. Meenakshi and M.S. Bhat have presented a robust control method based on H_2 controller for lateral stabilisation of a micro air vehicle. The developed methodology is applicable to all classes of mini and micro air vehicles.

The next two papers are on robot teams. H. Furukawa has designed an ecological interface for human supervision of a robot team. The performance of an embedded real-time system for use in a scenario of soccer playing robots has been evaluated by K. Koker.

The next two papers are on biped walking robot. Takahashi et al., have detailed the development of a simple six-axis biped robot system which can be used as a teaching aid for undergraduate students studying intelligent robotics. In the other paper the autonomous stride-frequency and step-length adjustment for biped walking control has been reported by L. Yang et al.

An extended Kalman filtering technique has been applied for the identification of unmanned aerial vehicle by A. G. Kallapur et al. M Goyal has reported a novel strategy for multi-agent coalitions in a dynamic hostile world.

K. H. Hyun et al., have designed a filter bank and extracted emotional speech features for the recognition of emotion based on voice. A FPGA-based implementation of graph colouring algorithms has been reported by V. Sklyarov and his colleagues.

O. Kenta et al., have employed levy based control strategy and extended it with autonomous role selection to solve the sharing problems of common resources.

R. Akmeliawati and I. M. Y. Mareels have studied the complex stability radius for automatic design of gain scheduled control in the next two papers.

In the last paper V. Sklyarov and I. Skliarov have suggested design methods for reconfigurable hierarchical finite state machines (RHFSM), which possess two important features such as they enable the control algorithms to be divided in modules providing direct support for "divide and conquer" strategy and they allow for static and dynamic reconfiguration.

We do hope that the readers will find this special issue interesting and useful in their research as well as in practical engineering work in the area of robotics and autonomous robots. We are very happy to be able to offer the readers such a diverse special issue, both in terms of its topical coverage and geographic representation.

Finally, we would like to whole-heartedly thank all the authors who have contributed their work to this special issue. We are grateful to all the reviewers from all around the world who have generously devoted their precious time to review the manuscripts.

Subhas Chandra Mukhopadhyay, Guest Editor
Institute of Information Sciences and Technology,
Massey University (Turitea Campus)
Palmerston North, New Zealand
S.C.Mukhopadhyay@massey.ac.nz

Gourab Sen Gupta, Guest Editor
School of Electrical and Electronic Engineering,
Singapore Polytechnic
Singapore
sengupta@sp.edu.sg

Dr. Subhas Chandra Mukhopadhyay graduated from the Department of Electrical Engineering, Jadavpur University, Calcutta, India in 1987 with a Gold medal and received the Master of Electrical Engineering degree from Indian Institute of Science, Bangalore, India in 1989. He obtained the PhD (Eng.) degree from Jadavpur University, India in 1994 and Doctor of Engineering degree from Kanazawa University, Japan in 2000.

During 1989-90 he worked almost 2 years in the research and development department of Crompton Greaves Ltd., India. In 1990 he joined as a Lecturer in the Electrical Engineering department, Jadavpur University, India and he promoted to Senior Lecturer of the same department in 1995.

Obtaining Monbusho fellowship he went to Japan in 1995. He worked with Kanazawa University, Japan as researcher and Assistant professor till September 2000.

In September 2000 he joined as Senior Lecturer in the Institute of Information Sciences and Technology, Massey University, New Zealand where he is working currently as an Associate professor. His fields of interest include Sensors and Sensing Technology, Electromagnetics, control, electrical machines and numerical field calculation etc.

He has authored 175 papers in different international journals and conferences, co-authored a book and written a book chapter and edited six conference proceedings. He has also edited two special issues of international journals (IEEE Sensors Journal and IJISTA) as guest editor.

He is a Fellow of IET (UK), a senior member of IEEE (USA), an associate editor of IEEE Sensors journal. He is in the editorial board of e-Journal on Non-Destructive Testing, Sensors and Transducers, Transactions on Systems, Signals and Devices (TSSD). He is in the technical programme committee of IEEE Sensors conference, IEEE IMTC conference and IEEE DELTA conference. He was the Technical Programme Chair of ICARA 2004 and ICARA 2006. He was the General chair of ICST 2005. He is organizing the ICST 2007 (icst.massey.ac.nz) as the General chair during November 26-28, 2007 at Palmerston North, New Zealand.

Mr. Gourab Sen Gupta graduated with a Bachelor of Engineering (Electronics) from the University of Indore, India in 1982. In 1984 he got his Master of Electronics Engineering from the University of Eindhoven, The Netherlands. In 1984 he joined Philips India and worked as an Automation Engineer in the consumer electronics division till 1989. Currently he works as a Senior Lecturer in the School of Electrical and Electronic Engineering at Singapore Polytechnic, Singapore. He was with Massey University, Palmerston

North, New Zealand from September 2002 to August 2006 as a Visiting Senior Lecturer while carrying out studies for his PhD degree.

He has published over 50 papers in international journals and conference proceedings, co-authored two books on programming, edited four conference proceedings, and edited two special issues of international journals (IEEE Sensors Journal and IJISTA) as guest editor. He is a senior member of IEEE. His area of interest is robotics, autonomous systems, embedded controllers and vision processing for real-time applications. He was the General Chair of ICARA 2004 and ICARA 2006. He was the Technical Programme chair of ICST 2005. He is organizing the ICST 2007 (icst.massey.ac.nz) as the Technical Programme Chair during November 26-28, 2007 at Palmerston North, New Zealand.

Contents

1

Toward Hierarchical Multi-Robot Urban Search and Rescue: Development of a 'Mother' Agent

David A. Williamson and Dale A. Carnegie

School of Chemical and Physical Sciences
Victoria University of Wellington, Wellington
New Zealand
david.williamson@vuw.ac.nz, dale.carnegie@vuw.ac.nz

Summary. This chapter describes the development of an embedded platform constituting the core of an Urban Search and Rescue (USAR) system. High level components include an embedded computer, network camera, supervisory module and wireless gateway. A distributed high efficiency power supply complements the low consumption of system components. Innovative design solutions have resulted in a compact, inexpensive and efficient embedded platform for robotic USAR.

Keywords: USAR, search and rescue, robot, embedded, ARM7.

1.1 Introduction

Robotic-assisted Urban Search and Rescue (USAR) aims to minimise human involvement in high-risk emergency situations. The efficiency of reported standalone SAR robots has been impeded by several factors including accessibility, efficiency and cost [1, 2]. A three-layer hierarchical system is being developed to address these issues [3]. The system in development expands on reported two-level systems [4–6] by insisting that the additional lowest level 'Daughter' robots are disposable.

A solution to the central layer 'Mother' robot is being developed. Mother robots are responsible for enabling and managing core system functions. Subsequent sections offer a complete system overview followed by details on the design and testing of embedded hardware.

1.2 System Design

Functional requirements of a Mother robot include high mobility, semi-autonomous operation, Daughter supervision and manual override. Design restrictions include cost, efficiency, size, weight and run time.

D.A. Williamson and D.A. Carnegie: *Toward Hierarchical Multi-Robot Urban Search and Rescue: Development of a 'Mother' Agent*, Studies in Computational Intelligence (SCI) **76**, 1–7 (2007)
www.springerlink.com

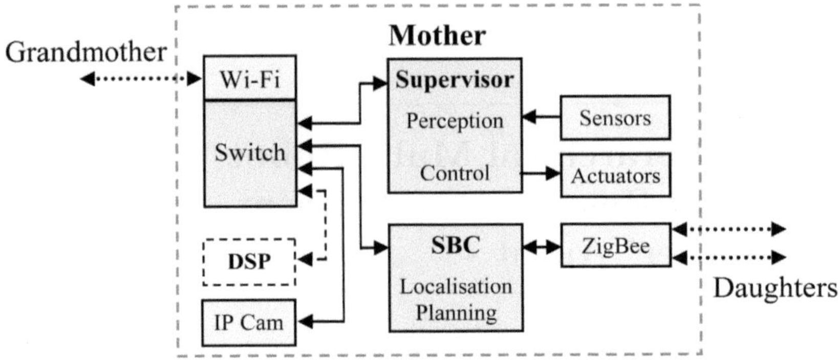

Fig. 1.1. High-level Mother robot hardware design

Figure 1.1 provides a high-level overview of the system in which a Wi-Fi router connects network-based system components to a top level 'Grandmother' robot. These include a single board computer, IP camera and a supervisory module. The supervisory module facilitates acquisition and control of sensors and actuators. An unused ethernet port facilitates hardware expansion if required for enhanced image processing.

Communication between Mother and Grandmother robots is based on IEEE 802.11g. Appreciable ranges are possible using high-power transceivers at data rates below 10% of the 54 Mbps link capacity. Core Mother components emulate host functionality to facilitate access by multiple users to live data and A/V streams. Separate control and observer access profiles for each resource prevent system contention.

The IEEE 802.15.4a based ZigBee communications standard has been nominated for Mother to Daughter communications. 802.15.4a adds precision ranging capability to the base 802.15.4 standard and is optimised for low power, low cost and medium range (100 m) applications. The 260 kbps offered by current 2.4 GHz ISM modules is sufficient for low resolution images or audio.

An AMD Geode based iEi WAFER-LX single board computer (SBC) provides the Mother with 800 MHz of processing power at 6 W. Prior experience within the VUW Mechatronics Group [7] suggests this will be sufficient to implement all of Mother's required high-level functionality. The Damn Small Linux-Not operating system has been installed on a high speed 1 GB compact flash memory card.

A network camera serves an audio–visual stream directly to multiple remote viewers. The local SBC grabs and process still images as required. The AXIS 207 IP network camera selected features 30-fps full VGA MPEG-4 encoding, achieving up to ten times the compression of MJPEG. This high efficiency codec increases effective WLAN operating range.

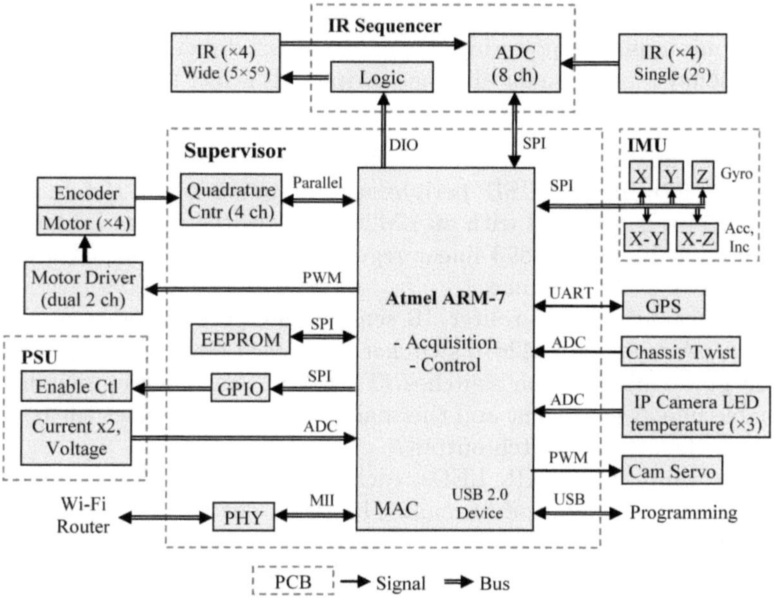

Fig. 1.2. Supervisor/acquisition and control module diagram

1.3 Acquisition and Control

The Mother SBC lacks the low-level peripheral interfaces required by the selected sensors and actuators. The system also requires a supervisor to control power distribution, administer override commands and reboot the SBC. These requirements are met by an Atmel AT91SAM7X ARM7 microcontroller. Figure 1.2 provides an overview of the sensors and actuators interfaced to the AT91SAM7X. The ethernet media access controller (MAC) and physical layer transceiver (PHY)(DP83848C) interface the module to high-level system components. Hardware resource requirements and loading by high-frequency interrupts due to the incremental encoders on each wheel drive motor is minimised by a 24-bit four axis quadrature decoder IC (LSI7566).

Motion feedback is provided by a MEMs based six degrees-of-freedom IMU. Two ADIS16201 dual-axis accelerometers provide both 14-bit acceleration and 12-bit inclination outputs. Three ADIS16080 gyros provide 12-bit angular rate outputs. Each sensor is digitally interfaced and aligned using an orthogonal PCB arrangement.

1.4 Power Supply

Estimates for operating power consumption yield 20 W for electronic hardware and 80 W for motors. One hour travelling and 2 h stationary operation would require a minimum 140 W h of battery power. Three 4 000 mA h 4-cell series lithium polymer (Li–Po) packs rated at 14.8 V provide 178 W h. Regulation losses are minimised by powering the motor drivers directly from the batteries. High efficiency regulation for all other peripherals is met through a distributed supply with the following facets:

1. A 15 W capacity supply to meet the 6 W at 5 V Wafer-LX SBC requirements, allowing for USB peripheral and possible PC-104 bus current draw. This is achieved with an LM2677-adj switch mode regulator. Post-regulation by an LP3853 linear regulator ensures an output well within the SBC's ±5% tolerance.
2. The IP camera, Wi-Fi router, IR sensors and wheel encoders also operate on 5 V. A separate LM2670 switcher feeds each device through four high-side power distribution switches (TPS20xx). Each power switch has an enable line, over current and thermal protection. A camera tilt R/C servo shares the camera switch output.
3. Three white Cree XR LEDs each provide 47 lum W^{-1} illumination to improve camera performance in dark environments. An LM2670 adjustable switcher supplies each LED up to 700 mA though 1 A power distribution switches for flexible luminescence and power control.
4. The AT91SAM7X acquisition and control module operates primarily from 3.3 V, with 5 V required only for the IMU gyros and level conversion logic. An LM2672-adj switcher is post-regulated by a linear LP3855 to provide 3.3 V at 1 A with remote sense feedback. An LM9076 provides 5 V at 150 mA.

The LM267x family's 260-kHz internal clock frequency allows use of moderate to small components while still achieving high efficiencies. Switcher outputs have been closely matched to the maximum dropout voltages of the LP385x linear regulators to maintain efficiency.

Two measures have been implemented to minimise noise induced by cross coupling between switching supplies. Each switcher is synchronised to an LMC555 derived 280 kHz clock. Parallel damped LC input filters optimised for minimum peaking have also been added before each LM267x [8]. This filter is advantageous in that damping requirements are simpler to achieve than for a pure LC filter.

1.4.1 Current and Voltage Monitor

Two high-side current sense monitors (ZXCT1010) are used to monitor the consumption of both the motors and all other hardware. The current senses are calibrated to provide a 1 V output at a maximum expected current of 6.5 A and 40 A, respectively. The sense outputs are amplified by 3.3 to match the AT91SAM7X ADC, which subsequently returns resolutions of 6.3 and 39 mA per LSB, respectively.

The AT91SAM7X also monitors battery voltage. The raw change between charged and discharged battery states is 4.8 V. A 10 V precision reference is employed to reduce the 12 V DC offset of the Li–Po packs. The resulting change observed at the base of the reference zener is divided by a factor of two before entering the ADC to provide a resolution of 6.44 mV per LSB.

1.4.2 Power Management and Shutdown

Eight peripherals including the SBC, IP camera, router, IR sensors, wheel encoders and three LEDs each feature logic enabled control with a manual override. The circuit implemented automatically enables an associated supply when one or more dependent power distribution switches are enabled. Manual override precedes AT91SAM7X control.

The AT91SAM7X can prioritise peripheral usage based on estimated remaining capacity. Over discharge protection is provided through an ICL7665 voltage monitor with a cut-off of 3.25 V and 0.25 V of hysteresis per cell. AT91SAM7X power is only disabled through a low battery voltage condition. Under-voltage shutdown consumption is 15 mA.

1.5 Discussion

The hardware described in the above sections has been constructed, tested and integrated to form a functional proof-of-concept Mother robot. The platform utilised is four-wheel drive and steered via differential skid. Suspension is facilitated by passive longitudinal articulation. Chassis construction is from welded aluminium and weighs 8.8 kg. Overall dimensions are $0.4 \times 0.5 \times 0.1$ m ($l \times w \times h$) while each wheel measures 0.3 m diameter.

Field trials on mobility have been conducted in an environment highly representative of urban debris as shown in Fig. 1.3. A ground clearance of 0.1 m, when combined with an optimal trajectory, permitted navigating obstacles up to 0.2 m high. The average forward velocity in such environments was observed to be $0.14\,\mathrm{m\,s^{-1}}$. Peak attainable velocity is $0.45\,\mathrm{m\,s^{-1}}$.

Net power consumption of the Mother platform has been measured against both incline and velocity. The results are summarised in Fig. 1.4 where all

Fig. 1.3. Mother chassis during field mobility trial

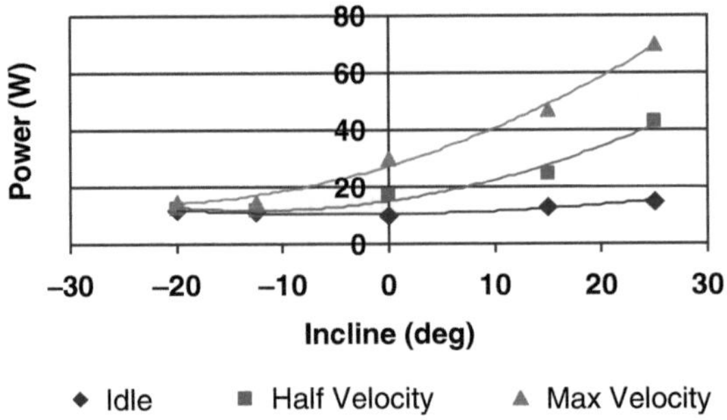

Fig. 1.4. Power consumption vs. grade

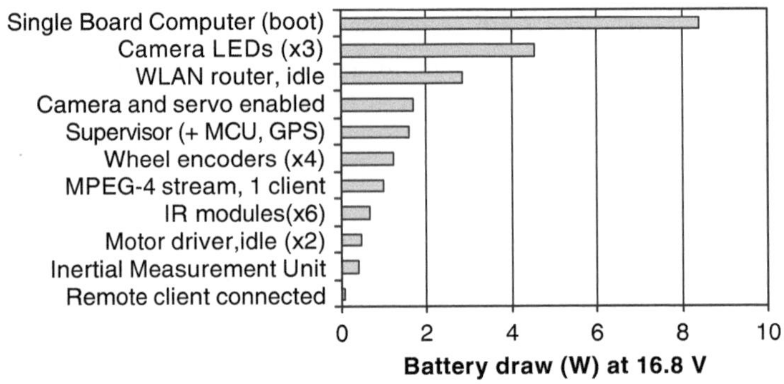

Fig. 1.5. Decomposition of hardware power consumption

peripherals except the SBC and camera LEDs were active. Operation on precipitous terrain consumes approximately 50 W at half velocity, indicating a capacity of 1.2 h or 0.9 km travel per battery pack.

Figure 1.5 provides a decomposition of the power drawn by various hardware components when measured in series with the batteries. The minimum power required to sustain idle control is 5.4 W, providing around 11 h operation per battery pack. Enabling only peripherals critical to tele-operation (no LEDs or SBC) immediately doubles power consumption.

Trials on wireless (802.11 g) communications indicate connections can be sustained at ranges greater than 30 m indoors and 300 m outdoors (line-of-sight) at the 5 Mbps data rate required for tele-operation. These distances are well within the platform's capacity, providing a substantial window of operating time in which Daughter robots could be deployed and controlled.

Present modes of Mother operation include full or shared manual control via joystick interfaced to a remote PC. Shared control fuses user intent with an IR-based Minimal Vector Field Histogram implementation to assist with obstacle avoidance [9].

1.6 Summary

This chapter presents an innovative approach to the design of a compact, low-power, inexpensive semi-autonomous robot. The authors believe that the approach used in this design will be of substantial interest to researchers wishing to construct a similar scale device. The USAR application of this design has necessitated the integration of a number of sensors, communications devices and mechanical ruggedness, which may be of particular interest to more specialist robotic developers.

A functional Mother chassis, complete with embedded components, has been designed, assembled and tested. Compliance with design requirements is satisfactory for this proof-of-concept system.

References

1. Drenner A, Burt I, Kratochvil B, Nelson BJ, Papanikolopoulos N, Yesin KB (2002) Communication and mobility enhancements to the scout robot. In: Proc of the IEEE/RSJ Int'l Conference on Intelligent Robots and Systems, Lausanne, Switzerland. Vol 1, pp 865–870
2. Yim MH, Duff DG, Roufas KD (200) Modular reconfigurable robots, an approach to urban search and rescue. In: Proc of the First International Workshop on Human Welfare Robotics Systems, Taejon, Korea. pp 19–20
3. Williamson DA, Carnegie DA (2006) Embedded Platform for Urban Search and Rescue Applications. In: Proc of the 3rd Int'l Conference on Autonomous Robots and Agents, Palmerston North, New Zealand, pp 373–378
4. Murphy RR, Ausmus M, Bugajska M, Ellis T, Johnson T, Kelley N, Kiefer J, Pollock L (1999) Marsupial-like Mobile Robot Societies. In: Proc of Autonomous Agents, Seattle, USA, pp 364–365
5. Jackson B, Burt I, Kratochvil BE, Papanikolopoulos N (2003) A Control System for Teams of Off-the-Shelf Scouts and MegaScouts. In: Proc of the 11th IEEE Mediterranean Conference on Control and Automation, Rhodes, Greece.
6. P.E. Rybski, N.P. Papanikolopoulos, S.A. Stoeter, D.G. Krantz, K.B. Yesin, M. Gini, R. Voyles, D.F. Hougen, B. Nelson, M.D. Erickson (2000) Enlisting Rangers and Scouts for Reconnaissance and Surveillance. In: IEEE Robotics & Automation Magazine, vol 7, Issue 4, pp 14–24
7. Carnegie DA (2002) Towards a Fleet of Autonomous Mobile Mechatrons. In: Proc of the 2nd Int'l Federation of Automatic Control Conference on Mechatronics Systems, California, USA, pp 673–678
8. Sclocchi M (2001) Input Filter Design for Switching Power Supplies, National Semiconductor Corporation Application Note
9. Levine SP, Bell DA, Jaros LA, Simpson RC, Koren Y, Borenstein J (1999) The NavChair Assistive Wheelchair Navigation System. In: IEEE Transactions on Rehabilitation Engineering, Vol 7, No. 4, pp 443–451

2

Multi-Robot Search and Rescue: A Potential Field Based Approach

J.L. Baxter[1], E.K. Burke[1]. J.M. Garibaldi[1], and M. Norman[2]

[1] Automated Scheduling, Optimisation and Planning Group, School of Computer Science and Information Technology, University of Nottingham, Nottingham, England. jlb@cs.nott.ac.uk, ekb@cs.nott.ac.uk, jmg@cs.nott.ac.uk

[2] Merlin Systems Corp Ltd, Plymouth International Business Park, Plymouth, England. marknorman@merlinsystemscorp.co.uk

Summary. This paper describes two implementations of a potential field sharing multi-robot system which we term as pessimistic and optimistic. Unlike other multi-robot systems in which coordination is designed explicitly, it is an emergent property of our system. The robots perform no reasoning and are purely reactive in nature. We extend our previous work in simulated search and rescue where there was only one target to the search for multiple targets. As in our previous work the sharing systems with six or more robots outperformed the equivalent non-sharing system. We conclude that potential field sharing has a positive impact on robots involved in a search and rescue problem.

Keywords: potential fields, multi-robot systems, search and rescue.

2.1 Introduction

Potential fields [11] have been used in robot navigation [17] for a number of years, despite a number of well known issues [13]. An example of such a difficulty includes oscillation near obstacles and in narrow passages. Modifications to the potential field algorithm have been proposed [19] to overcome these challenges. Other approaches to potential fields include that of Reif et al. [18] in which an individual agent's motion is a result of an artificial force imposed by other agents and components of the system. Damas et al. [7] modified potential fields to enhance the relevance of obstacles in the direction of the robot's motion. Howard et al. [10] divided their potential field into two components, a field due to obstacles and a field due to other robots. Pathak et al. [17] stabilised their robot within a surrounding circular area ('bubble') using two potential field controllers. The ROBOT was centred within a bubble and then its orientation was corrected.

In our approach, a robot's motion is a result of the force imposed by obstacles. In addition, a local group of robots share information on common

J.L. Baxter et al.: *Multi-Robot Search and Rescue: A Potential Field Based Approach*, Studies in Computational Intelligence (SCI) **76**, 9–16 (2007)
www.springerlink.com

potential field regions so that a robot's motion can be a result of obstacles not perceived by the individual robot. The concept of a local group is similar to that of dynamic robot networks [5]. However, instead of broadcasting trajectories and plans of robots, in our system the potential field information is broadcast. To the best of our knowledge, no previous research using potential fields has incorporated the concept of sharing potential field information amongst robots.

Unlike perception–reasoning–execution architectures [14] the system presented in this chapter does not reason about its environment. We describe our system as a reactive system [1]. Motion is based purely on the potential field created by the sensor data in real-time. There is no concept of teamwork [6] or role selection [20] in that each robot performs actions as an individual. Robots are not aware that they are part of a collective; co-ordination becomes an emergent property of the system through the implicit sharing of potential fields by the robots.

In this chapter, we extend our previous work [2] by using a potential field based architecture to search for two targets in an unknown environment. We largely reproduce the problem description in Sect. 2 from [2] for completeness. Although this is a specific problem instance, we believe our system may be applied to many other problems such as robotic soccer [7, 12], the coverage problem [10] and robotic hunting [3].

In the rest of this chapter, Sect. 2.2 describes the potential field based implementation, Sect. 2.3 describes the experiments undertaken, Sect. 2.4 outlines the results from the experiments and, finally, Sect. 2.5 is a discussion of the results observed and possible future work.

2.2 Potential Field Implementation

Before a description of the sharing robots is given, the individual robot system will be summarised, as the sharing robots are based upon this individual robot system and they are compared against it during the experimentation described in Sect. 2.3.

The individual robot system is made up of a number of robots which attempt to solve a task individually without any communication or co-ordination with one another. The system is implemented using a model based upon Coulomb's law of electrostatic force (as described in [4]). The system is composed of positively charged particles, which are used to calculate the field by the inverse square law below (2.1):

$$F = \frac{q_1 q_2}{r^2} \tag{2.1}$$

where q_1 and q_2 represent the charges of two particles and r is the distance between them. The resultant force, F, either repels or attracts the particles to one another. Using (2.1), we calculate eight individual forces, $f1 - f8$; r is

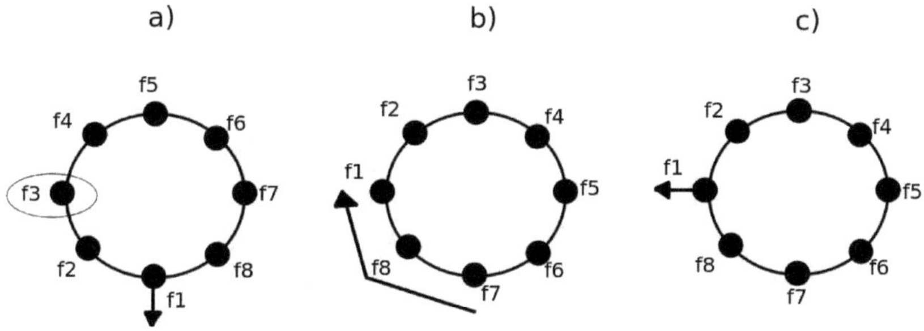

Fig. 2.1. Action selection: (**a**) the robot calculates the minimum force (f3), (**b**) the robot rotates towards the minimum force, (**c**) the robot moves forwards (towards the minimum force)

the range to an object (in metres) obtained from an ultrasonic sensor reading (see Fig. 2.1a), q_1 is the charge of the robot and q_2 is the charge of the object. For simplicity, all objects are represented by a positive unit charge.

Rather than formally resolving into a single force, the robot's motor control is a simple action selection of either *move forward* or *rotate*. When moving forward, the speed of the robot is proportional to the force acting upon it. When rotating, the angular speed is $0.5\,\mathrm{rad\,s^{-1}}$ and the forward speed is $0.025\,\mathrm{m\,s^{-1}}$. The robot calculates the minimum F (F_{min}) and rotates in the direction of F_{min}. If the direction of the robot equals the direction of F_{min} then the robot moves forwards (see Fig. 2.1).

Using this conceptually simple algorithm, the robot moves away from areas of positive charge (obstacles). As a target is indistinguishable from an obstacle to an ultrasonic sensor, a camera is used to differentiate between obstacles and targets using blob detection (targets are non-black obstacles). However, rather than giving the target a negative unit charge (-1), the task is said to be complete once the target has been found (this is a simplification due to the nature of the task described in Sect. 2.3). The orientation of the camera is fixed to the forward orientation of the robot (see Fig. 2.1a).

2.2.1 Potential Field Sharing

As noted earlier, the sharing robots are identical to the individual robots but implicitly share their potential field information with other robots within a local group. Therefore, by knowing only the relative positions (based upon odometric readings and the initial location of all robots) of other robots in the system, robots can assign themselves to a local group (the range of the local group catchment radius is set to an arbitrary value).

The world is represented as a two-dimensional plane. Simple geometric calculations (the intersection of circles and the intersection of lines and circles) are used to decide which of the ultrasonic sensors are likely (see Fig. 2.2a) to share information.

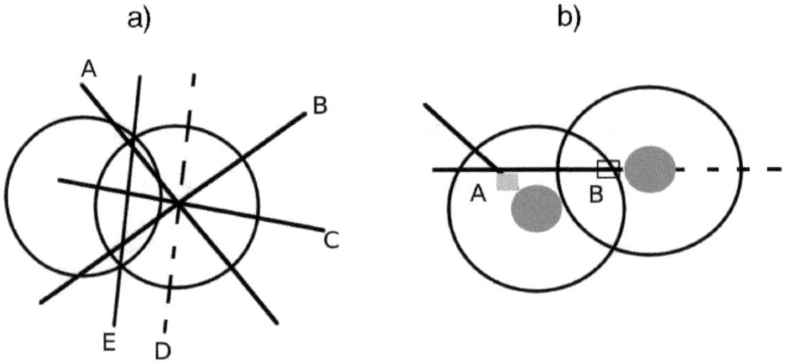

Fig. 2.2. (**a**) Likely sharing: The *circles* represent the local catchment areas of each robot. *Lines* A, B, C and D are the lines representing the ultrasonic sensors. Line E is the line intersecting the points of intersection of the two circles. (**b**) Two robots: The large circles represent the range of the robot's sensors and thus the *grey square* is an obstacle only observed by robot A. The *white square* represents what robot B would 'see' if the pessimistic system was implemented. The *dashed line* represents potential field information that is not shared

If any of these lines intersect, the line joining the points of intersection of each robot's local group catchment radius (line E in Fig. 2.2a). The intersections of lines are used to determine which lines (representing the direction of ultrasonic sensors) from any robot (within the local group radius) intersect any of the lines of another robot. Then the potential field information of the involved robot's ultrasonic sensors is shared. An example is given in Fig. 2.2b. Note that line D in Fig. 2.2a never needs to be considered, since it is parallel to line E. This process is repeated for all robots within the local catchment radius.

We describe two types of sharing systems. They are referred to as *pessimistic* and *optimistic* in the context of sensor noise. Consider, for example, the situation as in Fig. 2.2b in which robot A detects an obstacle that is not within robot B's sensor range. As they both belong to the same local group, a shared potential field is constructed. Robot A would suggest a high charge, whereas robot B would suggest a low charge. In the pessimistic system, the highest charge is selected and so an obstacle outside its own sensor range repulses robot B. In the optimistic system, the lowest charge is selected and thus the obstacle initially detected by its sensors no longer repulses robot A.

The advantage of the pessimistic system is that robots are less vulnerable to false negatives (and so avoid obstacles that they have not detected due to sensor noise), but the disadvantage is that they are more susceptible to false positives (and so avoid obstacles they do not need to). The advantage of the optimistic system is that robots are less vulnerable to false positives but are more susceptible to false negatives. Note that not all potential field information is shared amongst robots within the same local group, see Fig. 2.2b.

2.3 Experiments

As with our previous work [2], all experiments were conducted using the player/stage [8] simulator and the robots modeled in the simulator were Miabot Pros [15], with an ultrasonic range finder and avr-cam modules. The ultrasonic range finder is composed of eight ultrasonic sensors that give a 360° field of view with a range of 3 cm^{-1} m. The avr-cam has a field of view of 30° and can track up to eight blobs at the same time. The local group catchment radius is set to double the sonar range of 1 m. The simulated environment has approximately the same dimensions ($5 \times 3\,\text{m}^2$) as our real robot arena. As an extension of our previous work [2], the simulated robots have the task of discovering two targets in an unknown environment. Once both targets are found, the task is complete. The environment consists of obstacles, robots, a deployment zone and targets. Obstacles were generated within the environment (world) randomly; the size of the obstacle, its position and its orientation all varied. The positions of the targets and the deployment zone for the robots were also generated randomly. However, these positions were fixed throughout the experiments. The only difference between each experimental run is the noise within the simulation in the form of bad data. For example, sonar passing through obstacles. The deployment zone is a position in the world that all the robots start in (evenly spaced 0.1 m from one another in concentric circles starting from this initial position, where obstacles allow).

For these experiments, three worlds were generated (see Fig. 2.3). For each world, all three of the systems (individual, pessimistic and optimistic)

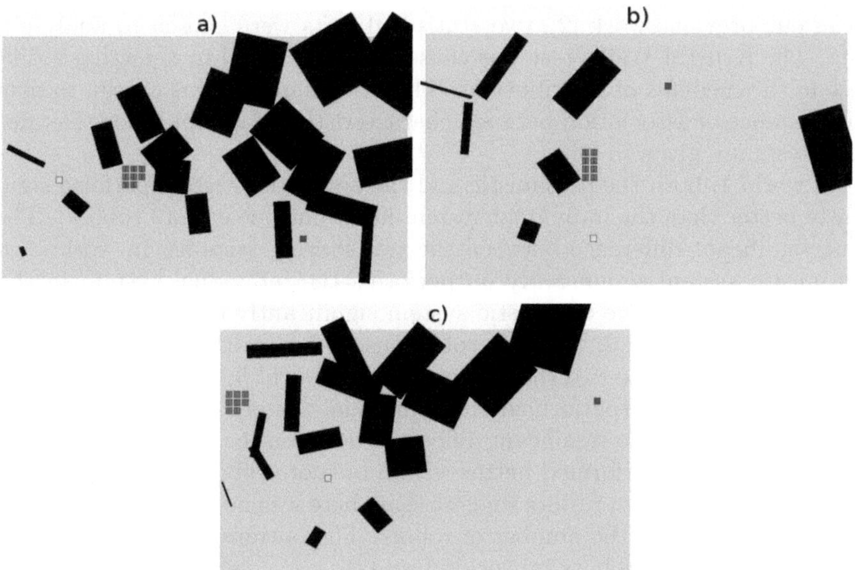

Fig. 2.3. (**a**) Simulated world 1. (**b**) Simulated world 2. (**c**) Simulated world 3. The group of eight squares are the simulated robots. The two lone squares are the targets. The *rectangles* are the obstacles

Table 2.1. Mean completion (seconds) for each system in each world for 2–8 robots, to one decimal place

	2	3	4	5	6	7	8
world 1							
ind	300.0	195.0	284.0	241.1	189.2	250.0	163.3
pes	184.5	183.5	104.8	135.1	126.3	165.0	104.0
opt	195.0	188.5	149.5	117.7	85.0	138.9	139.1
world 2							
ind	123.3	127.1	85.0	73.9	25.9	37.1	26.7
pes	70.2	48.5	53.8	62.3	37.7	28.0	25.4
opt	76.5	68.6	52.4	45.7	42.2	38.4	40.6
world 3							
ind	249.0	173.2	163.3	104.6	170.3	198.5	103.5
pes	116.2	94.5	135.7	114.5	98.8	103.6	83.3
opt	135.4	110.0	163.6	111.8	118.6	76.3	75.4

attempted 20 runs that were repeated for groups of 2–8 robots. The time recorded for both targets to be discovered was the metric recorded. Failure to complete the task within 300 s resulted in a score of 300 s. The means of the results are provided in Table 2.1.

2.4 Results

As in our previous work [2], two statistical tests were chosen to analyse the data. The Kruskal-Wallis test was chosen, as it is useful in detecting a difference in the medians of distributions. The Friedman test was chosen to detect the existence of association between characteristics of a population. Details of these tests are given in [9].

In world 1, both the pessimistic and the optimistic system perform significantly better than the individual system in all but one case (3 robots). There is no significant difference between the two sharing systems. In world 2, the pessimistic system significantly outperforms the individual system in the 2 and 3 robot cases. The optimistic system significantly outperforms the individual system in the 2, 3 and 5 robot cases. Again, there is no significant difference between the two sharing systems. In world 3, both the sharing systems significantly outperformed the individual system in the 2 and 7 robot cases and there was no significant difference between the sharing systems. The individual system performed better with 6 or more robots in worlds 1 and 2. The pessimistic system results suggest that there was no significant advantage gained by increasing the number of robots. The optimistic system performed best with 6 or more robots in worlds 1 and 2.

2.5 Discussion

The results show that the sharing systems perform significantly better than the individual system. The observation that there is no significant difference between the performance of the pessimistic and optimistic systems is also interesting as this implies that the ability to detect more obstacles has the same benefit as the ability to ignore more sonar noise. We believe that this may be due to the fact that there is a limited amount of noise within the simulation and the differences in the sharing systems' performance will be more apparent in the real world. The observation that both the individual and the optimistic system perform better with six or more robots makes sense as, in the case of the individual system, more robots in the world results in a greater area coverage. In the case of the optimistic system, as well as having the same benefits of the individual system, more robots result in more information being shared and so each robot can make better decisions. The observation that the pessimistic system did not benefit from an increased number of robots was not expected and requires further research.

A major limitation of the pessimistic and optimistic systems is their reliance upon accurate odometric readings. In the experiments carried out in this chapter, it was assumed that no errors occurred. In the real world, errors occur frequently due to wheel slippage. This will have to be accounted for in real world experiments.

Future work includes adapting the multi-robot systems to include group member recognition in order to improve robot dispersal. We also intend to investigate the effect of the local group radius (both increasing and decreasing its size). Other possible future work includes, applying this research to a very large-scale robotic system (hundreds of robots) [18] and implementing the sharing robots in other common problems such as robotic soccer [7, 12], formation control [16], the coverage problem [10] and hunting [3]. It is hoped that the work in this chapter can form a basis for future work on real robots.

References

1. R. Arkin. Reactive robotic systems. In M. Arbib, editor, *The Handbook of Brain Theory and Neural Networks*, pages 793–796. MIT Press, 1995.
2. J.L. Baxter, E.K. Burke, J.M. Garibaldi, and M. Norman. The effect of potential field sharing in multi-agent systems. In *The 3rd International Conference on Autonomous Robots and Agents (ICARA 2006)*, pages 33–38, Palmerston North, New Zealand, 12th-14th December 2006.
3. Z. Cao, M. Tan, L. Li, N. Gu, and S. Wang. Cooperative hunting by distributed mobile robots based on local interaction. *IEEE Transactions on Robotics*, 22(2):pages 403–407, 2006.
4. B.H. Chirgwin, C. Plumpton, and C. W. Kilmister. *Elementary Electromagnetic Theory*, volume 1. Pergamon Press, New York, 1971.
5. C.M. Clark, S.M. Rock, and J.-C. Latombe. Motion planning for multiple mobile robot systems using dynamic networks. In *IEEE Int. Conference on Robotics and Automation*, pages 4222–4227, 2003.

6. P.R. Cohen, H.J. Levesque, and I.A. Smith. On team formation. In J. Hintikka and R.Tuomela, editors, *Contemporary Action Theory, 2: Social Action*, pages 87–114. Kluwer, Dordrecht, 1997.
7. B.D. Damas, P.U. Lima, and L.M. Custdio. A modified potential fields method for robot navigation applied to dribbling in robotic soccer. In A. Kaminka R.R. Gal and P.U. Lima, editors, *RoboCup 2002: Robot Soccer World Cup VI*, pages 65–77. Springer Verlag, Berlin, 2003.
8. B. Gerkey, R.T. Vaughan, and A. Howard. The player/stage project: Tools for multi-robot and distributed sensor systems. In *Proceedings of the 11th International Conference on Advanced Robotics (ICAR 2003)*, pages 317–323, Coimbra, Portugal, 2003.
9. J. D. Gibbons. *Nonparametric Methods for Quantitative Analysis.* American Sciences Press, Columbus, Ohio, third edition, 1997.
10. A. Howard, M.J. Mataric, and G.S. Sukhatme. Mobile sensor network deployment using potential fields: A distributed, scalable solution to the area coverage problem. In *Distributed Autonomous Robotic Systems 5: Proceedings of the 6th International Symposium on Distributed Autonomous Robotics Systems(DARS02)*, pages 229–308, Fukuoka, Japan, 2002.
11. O. Khatib. Real-time 0bstacle avoidance for manipulators and mobile robots. In *IEEE Int. Conf. On Robotics and Automation*, pages 500–505, St. Loius, Missouri, 1985.
12. D.-H. Kim, Y.-J. Kim, K.-C. Kim, J.-H. Kim, and P. Vadakkepat. Vector field based path planning and petri-net based role selection mechanism with q-learning for soccer robots. *Int. J. Intelligent Automation and Soft Computing*, 6(1):pages 75–87, 2000.
13. Y. Koren and J. Borenstein. Potential field methods and their inherent limitations for mobile robot navigation. In *IEEE Conf. Robotics and Automation*, pages 1398–1404, 1991.
14. X.-W.T. Liu and J. Baltes. An intuitive and flexible architecture for intelligent mobile robots. In *The Second International Conference on Autonomous Robots and Agents(ICARA)*, pages 52–57, Palmerston North, New Zealand, 2004.
15. Merlin Systems Corporation. Ltd. Merlin robotics. http://www.merlinrobotics.co.uk.
16. S. Monterio and E. Bicho. A dynamical systems approach to behavior-based formation control. In *Int. Conf. Robotics Automation*, pages 2606–2611, Washington, 2002.
17. K. Pathak and S.K. Agrawal. An integrated path planning and control approach for nonholonomic unicycles using switched local potentials. *IEEE Transactions on Robotics*, 21(6):pages 1201–1208, 2005.
18. J. Reif and H. Wang. Social potential fields: A distributed behavioral control for autonomous robots. In R. Wilson K. Goldberg J.-C. Latombe and D. Halperin, editors, *International Workshop on Algorithmic Foundations of Robotics (WAFR)*, pages 431–459, Wellesley, MA, 1995. A. K. Peters.
19. J. Ren, K.A. McIssac, and R.V. Patel. Modified newtons method applied to potential field-based navigation for mobile robots. *IEEE Transactions on Robotics*, 22(2):pages 384–391, 2006.
20. H.L. Sng, G.S. Gupta, and C.H. Messom. Strategy for collaboration in robot soccer. In *The First IEEE International Workshop on Electronic Design, Test and Applications (DELTA 02)*, pages 347–351, Christchurch, 2002.

3

Probabilistic Target Search Strategy

R. Jarvis and M. Marzouqi

Intelligent Robotics Research Centre
Monash University, Melbourne
Australia
Ray.jarvis@eng.monash.edu.au

Summary. In this paper, a probabilistic based approach is presented by which a single mobile robotic agent might discover targets in an obstacle strewn environment in a strategic manner which minimizes the average time taken to find a target. The approach has been evaluated on simulated environments for the case of finding victims in a disaster area.

Keywords: Mobile robotics, target search strategy, distance transform based Gaussian distribution.

3.1 Introduction

Searching for a moving target in a cluttered known environment has been studied fairly extensively in the past, especially in the case of initially unknown target locations. In the work presented here [4], one or more targets are known or predicted to be at certain locations at some point in time before the search begins. Knowing the previous targets' locations can help in the search process where the notion of probabilities can be applied for an efficient search strategy. The solution is based on dividing the free spaces into convex regions and determining which regions should be searched first using a probabilistic measure. A recent work [1] presented a solution for a similar problem; however, both the probabilistic model of the targets' locations and the partitioning of the environment are supplied initially. In our case, these tasks are performed autonomously. The search path is then planned by finding the optimal order in which different convex regions are searched, where optimality is related to minimizing the average time for sighting a target.

Finding disaster victims will be studied here as one possible and important application. In a search and rescue situation over wide areas, perhaps initially known but possibly now devastated, it is clearly of value to find victims as quickly as possible, since their lives and limbs may depend on a timely rescue. But not everyone can be found first, so how best should a search for

R. Jarvis and M. Marzouqi: *Probabilistic Target Search Strategy*, Studies in Computational Intelligence (SCI) **76**, 17–23 (2007)
www.springerlink.com

victims be carried out? As a reasonable optimality criterion, minimizing the average time for sighting will take both numbers of victims and search times into consideration. Ultimately, diligence requires that the search extends to everywhere possible. However, the order in which different areas are searched will critically alter the optimality value calculated.

The next section outlines the assumptions made for the problem dealt with here. Next, the two main basic tools for providing solutions are described (two sections). The approach detail is described in Sect. 3.5. Results to date follow. Future developments are then suggested followed by conclusions.

3.2 Problem Assumptions

The assumption here is that only the finding (sighting) of victims is to be dealt with. Rescue after location is outside the scope of this chapter. It is to be admitted that many realistic details are not included in this study. These include the existing dangers to searchers, the practicality of entering various areas, visibility limitations in smoke, dust and fire or night conditions, whether victims can be communicated to by cell phone and so on. Thus this work is just a small beginning in thinking about search (followed by rescue) strategies.

Visibility is taken to mean the existence of a line of sight between the searcher and a victim. Only obstacles are considered as impediments to visibility, not smoke, dust, fire or other possibilities, such as size on the image plane. There exist algorithms for determining this kind of visibility in polyhedral obstacled space [7] and others that use simple ray tracings in rectangularly tessellated space [3]. Only two-dimensional spaces are considered here. This means that obstacles are projected on the flat ground plane. Three-dimensional extensions are not difficult to consider, but, nevertheless, will not be considered here.

A convex region in the plane is one which collects together (and finds the perimeter) spaces (or cells) within which every point is visible to every other point. Thus, it is adopted here that once a searcher enters any part of a convex region all parts of that region become visible and thus all victims in that region are considered found. If the initially known search environment were not altered, then visibility regions could be calculated 'a priori' with respect to the search. We will start with this unlikely situation and later consider what could be done about incrementally discovered changes. An alternative approach would be to consider that an aerial view has provided the knowledge of the environment after the disaster. Again, for simplicity, we will assume, initially, there is only one searcher and deal with multiple, co-operative searchers later.

3.3 Visibility-Based Map Segmentation

The map segmentation method [5] deals with stationary, obstacle strewn environments. It divides a tessellated free space into convex regions, i.e. the visibility line connecting any two cells within a region does not penetrate an obstacle. The map segmentation method is based on growing one convex region at a time from a chosen seed cell. The novelty of the method is to choose the next seed that has the lowest exposure to the free spaces amongst those 'unclassified cells' which do not yet belong to any previously created region. This result in creating a set of mostly logical partitions (e.g. rooms, hallways). The reason is that the chosen seed at any stage will be visible to unclassified cells that mostly belong to one separate physical space, which leads to a higher chance of forming a new convex region of that space. This will be clarified at the end of this subsection by an example. The map segmentation method is explained in [5].

3.4 Distance Transform Based Gaussian Distributions (DTGD)

One final component of the framework remains. This concerns the disposition of victims in terms of their spread amongst the convex regions of the environment. At least two factors relate to this aspect. The first is the expectation that knowledge of the type of activities likely to have been taken place in the various parts of the environment (e.g. street market, domestic activities, office operations, recreation concentrations, etc.) will allow some estimation of starting points of locations of victims and in what concentration. However, one would expect a spreading out after the disaster or perhaps during a disaster, due to panic and logic, but limited by obstacles, time and injuries so that some kind of spread function could be applied to the initial position/concentration estimates. In such situations, Gaussian distributions are favoured with standard deviations reflecting the spread extent. However, the spread is impeded by obstacles which cannot be penetrated. It is ludicrous to imagine concentrations along the edges of obstacles to reflect truncation of Gaussian functions. What we have discovered is a non-linear distance transform (DT) based value [6] which when substituted into the Gaussian exponent nicely models spreads favouring directions of accessibility in a realistic manner. The DT propagates cell step distances away from specified goals throughout all of reachable free space, flowing around the obstacles. The steepest descent path from any reachable free-space cell to the goal is optimal [2]. The method extends to higher dimensional spaces and can accommodate space/time considerations.

Details are to be found in (3.1) and (3.2) represents the modified Gaussian distribution function which calculates the probability at each free-space cell (x, y) given its already calculated transformed distance dxy. Normalizing the calculated probabilities is a necessary step to adjust their values to accurately sum to 1.0 (dividing each cell's value by the sum of all free-space cell values).

$$G_{DT}(x,y) = \frac{1}{2\pi\sigma^2} e^{-\frac{{d_{xy}}^2}{2\sigma^2}} \tag{3.1}$$

$$G_N(x,y) = \frac{G_{DT}(x,y)}{\sum_{x=0}^{x_{\max}} \sum_{y=0}^{y_{\max}} G_{DT}(x,y)} \tag{3.2}$$

3.5 Approach Details

The robot starts with a full or partial knowledge of the environment map and the initial locations of the victims. The search path is replanned if additional knowledge is acquired later. The details are as follow.

First, the probability map of the victim's current locations is constructed using the DTGD function. For each given location, the value of σ in (3.1) should be calculated based on the time elapsed since the victim was confirmed to be at its given location, and other related information such as its expected moving speed. σ has been set manually in the following experiments since estimating σ is not yet understood. The second step in the approach is to apply the map segmentation method to the environment map. This step can be performed simultaneously with the construction of the probability map.

The third step is to allocate a utility value for each region that represents the chance of sighting a victim there. The utility value of a region is the sum of the probability values calculated at each of its cells. Thus, a large convex region at the intersection of many routes can accumulate a high utility value. The search strategy aims to provide the order in which to visit all convex regions. Once a searcher robot enters any part of a region, all parts of that region become visible and thus all victims in that region are considered found. Therefore, the fourth step (which can be done simultaneously with the third step) is to choose a point within each region to be a destination node for the robot. Since it is unknown from which side the robot will enter each region, the centre of each region is chosen as the destination node. An exception is the region where the robot is initially located; the robot's location is this region's node. Using the DT algorithm, the distance between each node and other nodes are calculated and converted to travel time given the robot's speed.

The problem becomes a type of traveling salesman problem, except that a return to the start is not required and there are considerations other than minimal tour length to take into account. A reasonable optimality criterion to adopt is that of minimal average time for sighting, since this will take both numbers of victims and search time into consideration. Having the utility (U) of each node and the travel time (T) between each two nodes, the fifth step is to apply (3.3) which chooses amongst all possible visiting orders the one with the minimum average time to victim sighting. r is the number of nodes, where the first node in any order must be the robot's initial location. The average time to sighting of a certain visiting order is the sum of each node's utility multiplied by the total travel time to it; this sum is divided by the total utility of all nodes (a fixed value). A Genetic algorithm (GA) has been used to search for the optimal nodes visiting order, which can yield good but not necessarily optimal results. There are a number of ingenious ways of using a GA; the applied GA in the following experiments is described in [4].

$$\underset{all\ orders}{Min}\left[\frac{\sum_{i=2}^{r}\left(U_i \cdot \sum_{j=1}^{i-1} T_{j\rightarrow j+1}\right)}{\sum_{i=2}^{r} U_i}\right] \tag{3.3}$$

Finally, having chosen the best nodes visiting order, the robot starts the search by planning the shortest path (using the DT algorithm) to the next destination node. During navigation, the robot stops proceeding to the next node if: (1) it reaches the region's border of that node (i.e. the region has been viewed) or (2) it has already passed through the node's region earlier when approaching other nodes.

3.6 Results

Three experiments are shown in Fig. 3.1. Each experiment is represented by two parts. The first part shows the constructed probability distribution map of the targets' locations (σ is set to 40). The robot's start location is represented by 'R'. The victims' locations are represented by 'Target'. Obstacles are the dotted areas. A cell's brightness represents the sum of the Gaussian values at that cell for each victim location (a victim location represent a fixed number of people). The second part shows the created convex regions and the final search path. Each region has two values at its centre (node): the upper value is the visiting order of that region, while the lower value is the percentage of the region's utility value to the total utility value of all regions.

3.7 Discussion and Future Work

The spread of victims following a disaster clearly does not depend simply on spreading an estimated concentration of people in a structured way using a single value. What was happening before the disaster and the physical condition of victims immediately after the disaster and whether the disaster was progressive or global are all considerations worth including. However, many of these types of variations can be fairly easily be included by considering the time of day, events which may be starting finishing or in progress, the type of disaster (e.g. fire, earthquakes, nuclear attack, flood) and the extent of the damage to existing structures.

What has been presented here, despite the simplifying assumptions, is a good basis for further refining developments to determine an efficient search strategy. Future research could provide a more sophisticated modeling process

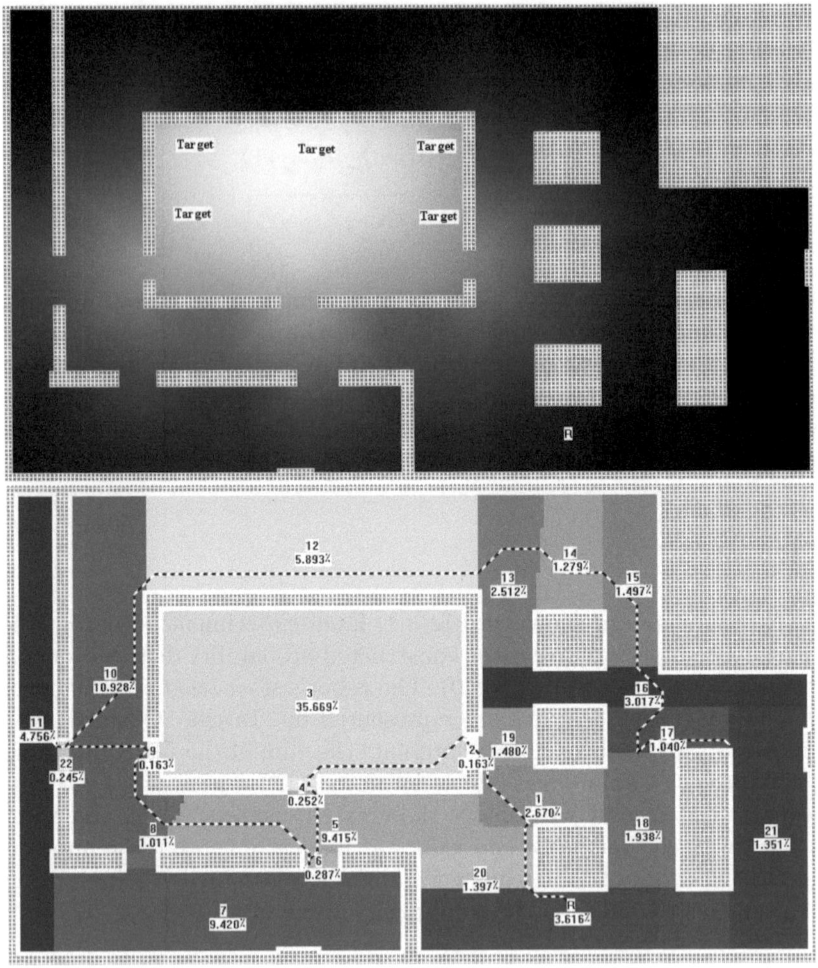

Fig. 3.1. A generated search path for disaster victims

that could better provide estimates of the numbers of people at various nominal points and the appropriate values to apply. The distance transform based Gaussian remains a very useful tool to accommodate obstacles in spread functions. The deployment of multiple robotic agents to work co-operatively in the rescue operation is also of interest.

3.8 Conclusion

This work has presented a means by which victim might be found after a disaster by a single robotic agent such that the average time to sight a victim is minimized. Many simplifying assumptions have been made but the notion of using a distance transform based Gaussian spread function, novel TSP based strategy of search order and the construction of 'visibility regions' is a sound basis for further development.

References

1. Lau H. Huang S. Dissanayake, G. (2005) Optimal search for multiple targets in a built environment. IEE/RSJ Int. Conf. on Intelligent Robots and Systems, pp. 3740–3745, August 2005.
2. Jarvis R. (1994) Distance Transform Based Collision-Free Path Planning for Robot Navigation in Known, Unknown and Time-Varying Environments. Invited chapter for a book entitled 'Recent Trends in Mobile Robots' edited by Professor Yuan F. Zheng World Scientific Publishing Co. Pty. Ltd. pp. 3–31.
3. Jarvis R. (1995) Collision-Free Trajectory Planning Using Distance Transforms. National Conference and Exhibition on Robotics, Melbourne, 20–24th Aug. 1984, also in Mechanical Engineering Transactions, Journal of the Institution of Engineers, Vol. ME10, No. 3, Sept. 1995, pp. 187–191.
4. Jarvis R., Marzouqi M. (2006) Efficient robotic search strategies for finding disaster victims. The 3rd International Conference on Autonomous Robots and Agents (ICARA), Palmerston North, New Zealand, December 2006.
5. Marzouqi M., Jarvis J. (2004) Efficient Robotic Pursuit of a Moving Target in a Known Environment Using a Novel Convex Region Segmentation. In proceedings of the Second International Conference on Autonomous Robots and Agents, pp: 13–18, 13–15th December, 2004, Massey University Palmerston North, New Zealand.
6. Marzouqi M., Jarvis R. (2005) The Dark Path Algorithm for Covert and Overt Path Planning with Uncertainties Involved. IEEE TENCON conference, Melbourne, 2005.
7. Munson J.H. (1971) Efficient Calculations on Points and lines in the Euclidean Plane. SRI AI Group Technical Note No. 61, Oct. 1971.

4

Localisation and Mapping With a Mobile Robot Using Sparse Range Data

Jochen Schmidt, Chee K. Wong, and Wai K. Yeap

Centre for Artificial Intelligence Research
Auckland University of Technology
Auckland, New Zealand
jochen.schmidt@aut.ac.nz, chee.wong@aut.ac.nz, wai.yeap@aut.ac.nz

Summary. We present an approach for indoor mapping and localisation using sparse range data, acquired by a mobile robot equipped with sonar sensors. The chapter consists of two main parts. First, a split and merge based method for dividing a given metric map into distinct regions is presented, thus creating a topological map in a metric framework. Spatial information extracted from this map is then used for self-localisation on the return home journey. The robot computes local confidence maps for two simple localisation strategies based on distance and relative orientation of regions. These local maps are then fused to produce overall confidence maps.

Keywords: Mobile robot, sonar, split and merge, data fusion.

4.1 Introduction

Mapping and self-localisation play an important role when using mobile robots for the exploration of an unknown environment. Particularly for indoor applications, where a 2-D map is usually sufficient, geometric maps obtained from time-of-flight devices, such as laser or sonar, are widely used. In this chapter, we present an algorithm for mapping and localisation using sparse range data acquired by only two sonars, and show that the robot can localise itself even with a map that is highly inaccurate in metric terms. In the first part of the chapter a method for dividing a given metric map into distinct regions, e.g. corridors or rooms is presented, thus creating a metric-topological map. As we use only two sonar sensors, the available range data are very sparse, therefore making the map highly inaccurate. We will show that these data can nevertheless be used for self-localisation. Our work is inspired by [11], where a cognitive map is regarded as a network of local spaces, each space described by its shape and its exits to other local spaces. Related approaches can be found, e.g. in [4], which is a hybrid approach that combines topological and

J. Schmidt et al.: *Localisation and Mapping With a Mobile Robot Using Sparse Range Data*, Studies in Computational Intelligence (SCI) **76**, 25–33 (2007)
www.springerlink.com

metric maps. In [8], topological maps are constructed from grid maps using Voronoi diagrams; the grid maps are split into regions and gateways are detected. In contrast to these methods, our approach is based on a region split and merge algorithm [5]. Many algorithms have been developed to solve the simultaneous mapping and localisation problem (SLAM) for mobile robots. For some examples of recent work in this area see [2] and the references therein. The approach followed in this chapter is different, as we do not solve the SLAM problem, but simulate a cognitive mapping process instead. The latter refers to the process in which humans and animals learn about their environment. We implemented two localisation strategies using both distance and orientation information extracted from the metric-topological map. We show how local confidence maps (location estimate of the robot) can be computed, and how to fuse them. For more details and experimental results refer to our earlier work [7, 10].

4.2 Mapping

The mapping process described in this section is used in two ways: first, the robot explores its environment and collects data. When this is finished, all acquired data are processed; the result of this initial mapping stage will be called the *original map* further on. This is the map the robot will use for returning home (origin of original map). An overview over the map-processing algorithm will be given in the following. A more detailed evaluation of the algorithm including the influence of the parameters involved is given in [6].

For data acquisition we use a mobile robot equipped with an odometer and two sonar sensors located on the left and right side. Note that the algorithms presented here are not restricted to sparse data or that type of sensors; the performance will be even better when more range data are available. The robot acquires sonar readings while moving on a 'straight' line (we are not concerned about drift compensation or correction) until it runs into an obstacle. At this point an obstacle avoidance algorithm is used, after which the robot can wander straight on again. One of these straight movements will be called *robot path* throughout this chapter. Based on the raw sonar sensor readings we build a geometric map containing the robot movement path as well as linear surfaces approximated from the sonar data. The goal is to split the map into distinct regions, e.g. corridors and rooms. Splitting is done along the robot movement path, using an objective function that computes the quality of a region, based on criteria such as the average room width (corridors are long and narrow compared to rooms) and overall direction (e.g. a corridor is separated from another one by a sharp bend in the wall). The basis of the map-processing algorithm is the well-known split and merge method [5]. In pattern recognition this algorithm is traditionally used for finding piecewise linear approximations of a set of contour points. Other applications include segmentation of image regions given a homogeneity criterion, e.g. with respect to colour or texture.

Before a region split and merge, algorithm on the geometric map can be applied, it is necessary to create an initial split of the map. The easiest way to do so is to treat the whole map as a single large region defined by the start and end points of the journey. After this step, the actual division of the map into distinct regions is performed based on a split and merge that uses a residual error function $g(\mathcal{S}_i, \mathcal{S}_j)$ which compares two regions $\mathcal{S}_i$ and $\mathcal{S}_j$ and computes the homogeneity of the two regions (low values of $g(\mathcal{S}_i, \mathcal{S}_j)$ means homogeneous, high values very inhomogeneous). This function is used during the split phase for deciding whether a region $\mathcal{S}_i^k$ will be split again at a given position into two new regions $\mathcal{S}_j^{k+1}$ and $\mathcal{S}_{j+1}^{k+1}$. If the homogeneity is above a given threshold θ_r, the region will be split again. When no further splitting is possible, the algorithm tries to merge adjacent regions (which were not necessarily generated by a single split) by checking whether the created region is still homogeneous. The basic idea is to use the average width of a region in the map as a criterion for splitting, as a width change resembles a changing environment, e.g. a transition from a corridor to a big room. The homogeneity (residual) function used is:

$$g(\mathcal{S}_i, \mathcal{S}_j) = \frac{\max\{f_\mathrm{w}(\mathcal{S}_i), f_\mathrm{w}(\mathcal{S}_j)\}}{\min\{f_\mathrm{w}(\mathcal{S}_i), f_\mathrm{w}(\mathcal{S}_j)\}} + s_\mathrm{r} r(\mathcal{S}_i, \mathcal{S}_j) \tag{4.1}$$

where $f_\mathrm{w}(\mathcal{S}_i)$ is the average width of region $\mathcal{S}_i$, and $r(\mathcal{S}_i, \mathcal{S}_j)$ is a regularisation term that takes care of additional constraints during splitting. The average width is given by $f_\mathrm{w}(\mathcal{S}_i) = \frac{A_{\mathcal{S}_i}}{l_{\mathcal{S}_i}}$, where $A_{\mathcal{S}_i}$ is the area of region $\mathcal{S}_i$, and $l_{\mathcal{S}_i}$ is its length. In practice, the computation of both needs a bit of attention. Particularly the definition of the length of a region is not always obvious, but can be handled using the robot movement paths, which are part of each region. The length $l_{\mathcal{S}_i}$ is then defined by the length of the line connecting the start point of the first robot path of a region and the end point of the last path of the region. This is a simple way to approximate a region's length without much disturbance caused by zig-zag movement of the robot during mapping.

Regarding the area computation, the gaps contained in the map have to be taken into account, either by closing all gaps or by using a fixed maximum distance for gaps. Both approaches have their advantages as well as drawbacks, e.g. closing a gap is good when it originated from missing sensor data, but may distort the splitting result when the gap is an actual part of the environment, thus enlarging a room. We decided to use a combined approach, i.e. small gaps are closed in a pre-processing step already, while large ones are treated as distant surfaces.

The regularisation term $r(\mathcal{S}_i, \mathcal{S}_j)$ ensures that the regions do not get too small. In contrast to a threshold, which is a clear decision, a regularisation term penalises small regions but still allows to create them if the overall quality is very good. We use a sigmoid function that can have values between -1 and 0, centred at n, which is the desired minimum size of a region:

$$r(\mathcal{S}_i, \mathcal{S}_j) = \frac{1}{1 + \exp\left(-\frac{\min\{A_{\mathcal{S}_i}, A_{\mathcal{S}_j}\}}{A_{\max}} + n\right)} - 1\,. \tag{4.2}$$

The exponent is basically the area of the smaller region in relation to the maximum area $A_{\max}$ of the smallest allowed region. This term only has an influence on small regions, making them less likely to be split again, while it has virtually no influence when the region is large, as the sigmoid reaches 0.

The influence of the term can be controlled using the factor s_{r} in (4.1), which is given by $s_{\mathrm{r}} = s\theta_{\mathrm{r}}$, where $0 \leq s \leq 1$ is set manually and defines the percentage of the threshold θ_{r} mentioned earlier that is to be used as a weight.

4.3 Localisation

Once the original map has been generated, we instruct the robot to return home based on the acquired map. Each time the robot stops on its return journey (because of an obstacle), it performs a map processing as described in Sect. 4.2, making a new metric-topological map available at each intermediate stop that can be used for localisation.

In the following, we describe two strategies for localisation, and a data fusion algorithm that allows for an overall position estimate computed from the individual localisation methods. Each method computes a local confidence map that contains a confidence value for each region of the original map. All local confidence maps are then fused into a single global one.

The fusion of local confidence maps, which may have been generated by different robot localisation methods with varying reliability, is based on the idea of *democratic integration* introduced in [9]. It was developed for the purpose of sensor data fusion in computer vision and computes confidence maps directly on images. The original method has been extended and embedded into a probabilistic framework in [1,3], still within the area of machine vision. We extend the original approach in a way that we do not use images as an input, but rather generate local confidence maps using various techniques for robot localisation. A main advantage of this approach is that the extension to more than two strategies is straightforward. Each local confidence map contains a confidence value between 0 and 1 for each region of the original map. As in [9] these confidence values are not probabilities, and they do not sum up to one; the interval has been chosen for convenience, and different intervals can be used as desired.

The actual fusion is straightforward, as it is done by computing a weighted sum of all local confidence maps. The main advantage of using democratic integration becomes visible only after that stage, when the weights get adjusted dynamically over time, depending on the reliabilities of the local map. Given M local confidence maps $\boldsymbol{c}_{\mathrm{loc}\,i}(t) \in \mathbb{R}^N$ (N being the total number of regions in the original map) at time t generated using different strategies, the global map $\boldsymbol{c}_{\mathrm{glob}}(t)$ is computed as $\boldsymbol{c}_{\mathrm{glob}}(t) = \sum_{i=0}^{M-1} w_i(t)\boldsymbol{c}_{\mathrm{loc}\,i}(t)$, where $w_i(t)$

are weighting factors that add up to one. An estimate of the current position of the robot with respect to the original map can now be computed by determining the largest confidence value in $\boldsymbol{c}_{\text{glob}}(t)$. Its position b in $\boldsymbol{c}_{\text{glob}}(t)$ is the index of the region that the robot believes it is in. The confidence value c_{glob_b} at that index gives an impression about how reliable the position estimate is in absolute terms, while comparing it to the other ones shows the reliability relative to other regions.

In order to update the weighting factors, the local confidence maps have to be normalised first; they are given by $\boldsymbol{c}'_{\text{loc}i}(t) = \frac{1}{N}\boldsymbol{c}_{\text{loc}i}(t)$. The idea when updating the weights is that local confidence maps that provide very reliable data get higher weights than those which are unreliable. Different ways for determining the quality of each local confidence map are presented in [9]. We use the normalised local confidence values at index b, which has been determined from the global confidence map as shown above, i.e. the quality $q_i(t)$ of each local map $\boldsymbol{c}_{\text{loc}i}(t)$ is given by $c'_{\text{loc}b}(t)$. Normalised qualities $q'_i(t)$ are computed by $q'_i(t) = \frac{q_i(t)}{\sum_{j=0}^{M-1} q_j(t)}$. The new weighting factors $w_i(t+1)$ can now be computed from the old ones: $w_i(t+1) = w_i(t) + \frac{1}{t+1}\left(q'_i(t) - w_i(t)\right)$. Using this updated equation and the normalisation of the qualities ensures that the sum of the weights equals one at all times [9].

Two strategies for computing local confidence maps are described below, one based on distance travelled, the other based on orientation information. Depending on the sensors used, more sophisticated ones can be added to enhance localisation accuracy. A main feature of data fusion is that each strategy taken on its own may be quite simple and not very useful for localisation; it is the *combination* of different strategies which makes localisation possible.

The first strategy is based on using the distance the robot travelled from its return point to the current position. Note that neither do we care about an exact measurement nor do we use the actual distance travelled as provided by odometry. Using the odometry data directly would result in very different distances for each journey, as the robot normally moves in a zig-zag fashion. Instead we use distance information computed from the region splitting of the maps, i.e. region length, which is defined by the distance between the 'entrance' and the 'exit' (split points) the robot used when passing through a particular region. The basic idea is to compare the distance d, measured in region lengths taken from the intermediate map computed on the return journey, to the lengths taken from the original map computed during the mapping process.

The confidence for each region in the local confidence map $\boldsymbol{c}_{\text{Dist}}$ depends on the overall distance d travelled on the return journey; the closer a region is to this distance from the origin, the more likely it is the one the robot is in currently. We decided to use a Gaussian to model the confidences for each region, the horizontal axis being the distance travelled in millimetre. The Gaussian is centred at the current overall distance travelled d. Its standard deviation σ is dependent on the distance travelled, and was chosen as $\sigma = 0.05d$. A Gaussian was chosen not to model a probability density, but for a number of reasons making it most suitable for our purpose: It allows for a smooth transition between regions, and the width can be easily adjusted by altering the standard deviation. This is necessary as the overall distance travelled gets more and more unreliable (due to slippage and drift) the farther the robot travels. The confidence value for a region is determined by sampling the Gaussian at the position given by the accumulated distances from the origin (i.e. where the robot started the homeward journey) to the end of this region. After a value for each region is computed, the local confidence map $\mathbf{c}_{\text{Dist}}$ is normalised to the interval $[0; 1]$.

The second strategy is based on using relative orientation information. We define the direction of a region as the direction of the line connecting the 'entrance' and 'exit'. Certainly this direction information varies every time the robot travels through the environment, but the overall shape between adjacent regions is relatively stable. Therefore, angles between region directions can be used as a measure of the current position of the robot. It has the advantage that angles between adjacent region directions are a local measure, thus keeping the influence of odometry errors to a minimum.

First, all angles $\alpha_1, \ldots, \alpha_{N-1}$ between adjacent regions in the original map are computed. In the re-mapping process while going home, new regions are computed in the new map based on data gathered while the robot travels. Using the direction information contained in this map, the angle β between the current region and the previous one can be computed. 'Comparing' this angle with all angles of the original map gives a clue (or many) for the current location of the robot, resulting in a local confidence map $\mathbf{c}_{\text{Dir}\,i} = \frac{1}{2}(\cos|\alpha_i - \beta| + 1)$. This results in high values for similar angles and low values for dissimilar ones.

4.4 Experimental Results

The main features of the office environment where we conducted the experiments are corridors, which open into bigger areas at certain locations, doors and obstacles like waste paper baskets in various positions. The acquisition of the original maps and the experiments for using the maps for localisation was done on different days, so the environment was different for each experiment (e.g. doors open/closed). We used an Activmedia Pioneer robot, equipped with an odometer and eight sonar sensors; only the two side sensors were used in order to obtain sparse range data.

Figure 4.1 shows four maps, including the locations of split points marked by dots. These are located on a set of connected lines that resemble the path the robot took while mapping the environment. To the left and right of that path, the (simplified) surfaces representing the environment can be seen. For splitting purposes, gaps were treated as distant surfaces, having a distance of 6 m from the position of the robot. The robot started the mapping process at the origin. All maps were processed using the same parameter values, $\theta_r = 2.0$ and $s = 0.1$; the desired minimum size of a region was 1.5 m. We found that the overall robustness to changes in the parameters is quite high, i.e. the choice of the actual values is usually noncritical; for an evaluation see [6]. It can be observed that the splits are located at the desired positions, i.e. where the environment changes, either from corridor to big room or at sharp bends in the corridor. The maps shown in Figs. 4.1a,b were generated from the mapping and going home processes, respectively, for Experiment 1; the Figs. 4.1c,d are the maps generated from the mapping and going home processes respectively for Experiment 2. Comparing the maps generated during mapping and going home highlights the difficulty in using these maps directly for localisation.

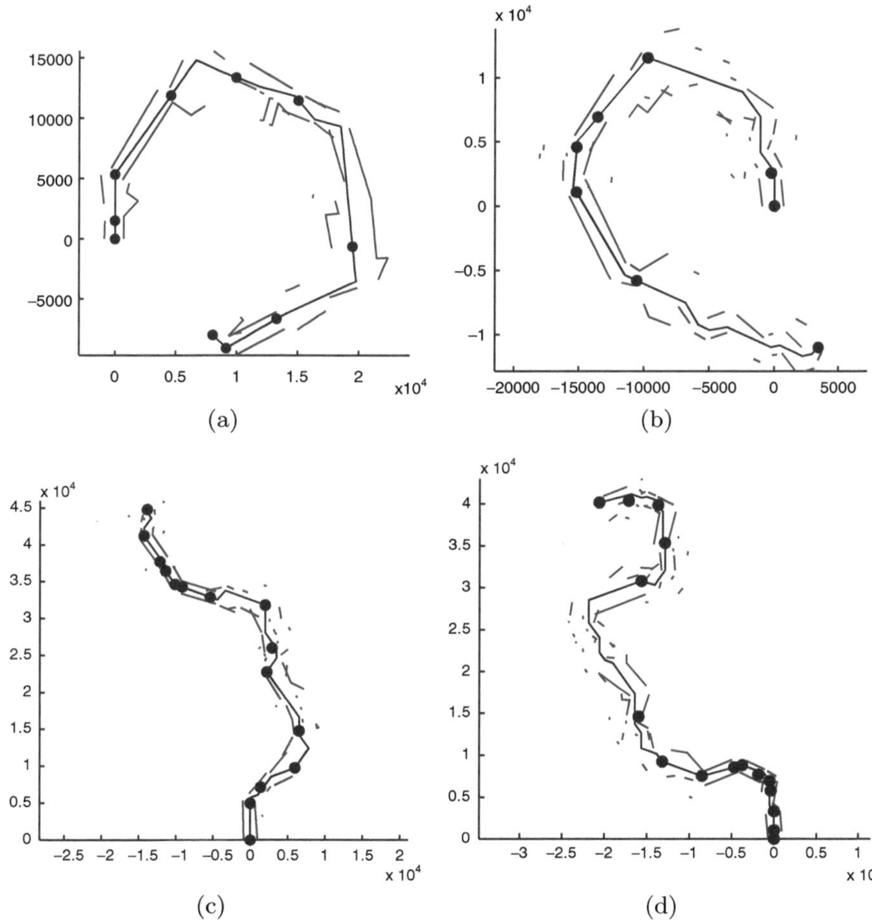

Fig. 4.1. (**a**) original map Experiment 1, (**b**) map generated during homeward journey Experiment 1, (**c**) original map Experiment 2, (**d**) homeward journey Experiment 2. *Black dots* indicate split points, robot movement starts at the origin

Each time, the robot goes through the same environment, it will generate different representations due to sensory inaccuracies.

Figure 4.2 shows two confidence maps for each experiment computed at different locations during the return home journey. The light dotted lines represent the region estimate using the region length information (distance method) and the dark dashed lines depict the region estimate using the angles between regions (relative orientation method). The solid line is the overall region estimate. The confidence maps in Fig. 4.2 illustrate different situations during localisation. A narrow peak for the overall confidence signifies the robot being very confident of being in a particular region. A wider confidence curve shows that the robot is at the transition from one region to another, as more than one region has a high confidence value, and the robot is unsure which of the regions it is in. Comparisons of the estimated position to the actual position have shown that the localisation is usually correct, with possible deviations of ± 1 in areas where the regions are extremely small.

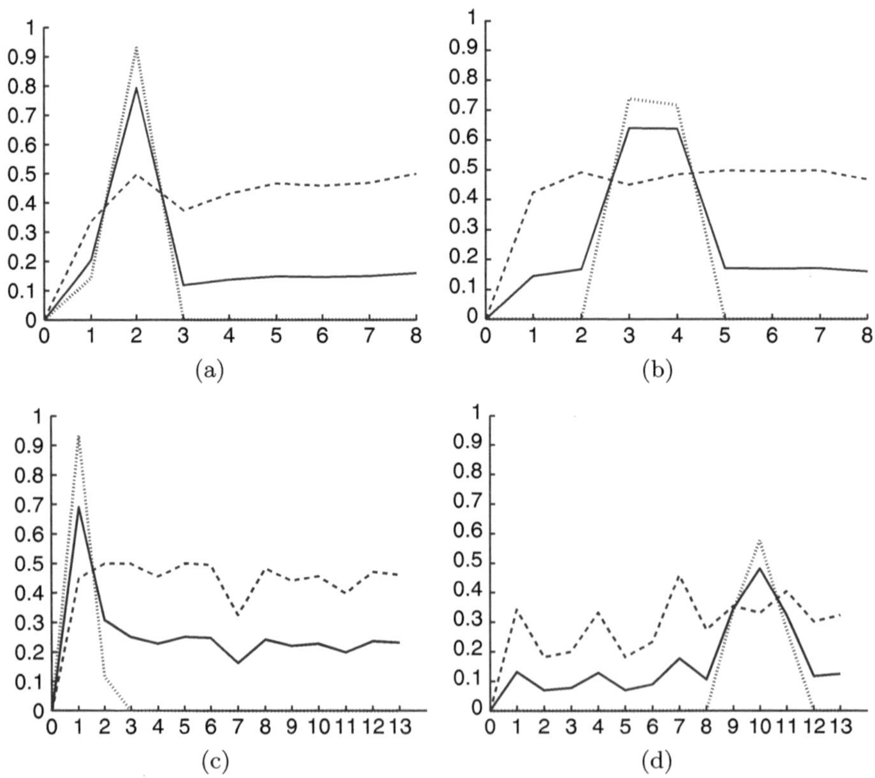

Fig. 4.2. Confidence maps at different locations. (**a**),(**b**) Experiment 1. (**c**),(**d**) Experiment 2. The plots show: distance (*light dotted*), relative orientation (*dark dashed*), overall confidence (*solid*). Horizontal axis: region index; vertical axis: confidence

4.5 Conclusion

We have presented methods for mapping and localisation using sparse range data. An initial metric map obtained from sonar sensor readings is divided into distinct regions, thus creating a metric-topological map. A split and merge approach has been used for this purpose, based on an objective function that computes the quality of a region. Based on spatial information derived from these maps, we showed how simple localisation strategies can be used to compute local confidence maps that are fused into a single global one, which reflects the confidence of the robot being in a particular region. The fusion can easily be extended by more localisation strategies or additional sensors.

References

1. J. Denzler, M. Zobel, and J. Triesch. Probabilistic Integration of Cues From Multiple Cameras. In R. Würtz, editor, *Dynamic Perception*, pages 309–314. Aka, Berlin, 2002.
2. C. Estrada, J. Neira, and J.D. Tardos. Hierarchical SLAM: Real-Time Accurate Mapping of Large Environments. *IEEE Trans. on Robotics*, 21(4):588–596, 2005.
3. O. Kähler, J. Denzler, and J. Triesch. Hierarchical Sensor Data Fusion by Probabilistic Cue Integration for Robust 3-D Object Tracking. In *IEEE Southwest Symp. on Image Analysis and Interpretation*, pages 216–220, Nevada, 2004.
4. B. Kuipers, J. Modayil, P. Beeson, M. MacMahon, and F. Savelli. Local Metrical and Global Topological Maps in the Hybrid Spatial Semantic Hierarchy. In *Int. Conf. on Robotics and Automation*, pages 4845–4851, New Orleans, LA, 2004.
5. T. Pavlidis and S.L. Horowitz. Segmentation of Plane Curves. *IEEE Trans. on Computers*, C-23:860 – 870, 1974.
6. J. Schmidt, C.K. Wong, and W.K. Yeap. A Split & Merge Approach to Metric-Topological Map-Building. In *Int. Conf. on Pattern Recognition (ICPR)*, volume 3, pages 1069–1072, Hong Kong, 2006.
7. J. Schmidt, C.K. Wong, and W.K. Yeap. Mapping and Localisation with Sparse Range Data. In S.C. Mukhopadhyay and G. Sen Gupta, editors, *Proceedings of the Third International Conference on Autonomous Robots and Agents (ICARA 2006)*, pages 497–502, Palmerston North, New Zealand, 2006.
8. S. Thrun. Learning Metric-Topological Maps for Indoor Mobile Robot Navigation. *Artificial Intelligence*, 99(1):21–71, 1998.
9. J. Triesch and Ch. von der Malsburg. Democratic Integration: Self-Organized Integration of Adaptive Cues. *Neural Computation*, 13(9):2049–2074, 2001.
10. C.K. Wong, J. Schmidt, and W.K. Yeap. Using a Mobile Robot for Cognitive Mapping. In *International Joint Conference on Artificial Intelligence (IJCAI)*, pages 2243–2248, Hyderabad, India, 2007.
11. W.K. Yeap and M.E. Jefferies. Computing a Representation of the Local Environment. *Artificial Intelligence*, 107(2):265–301, 1999.

5

Applying High-Level Understanding to Visual Localisation for Mapping

Trevor Taylor

Faculty of IT Queensland University of Technology
Brisbane, Australia

Summary. Digital cameras are often used on robots these days. One of the common limitations of these cameras is a relatively small field of view. Consequently, the camera is usually tilted downwards to see the floor immediately in front of the robot in order to avoid obstacles. With the camera tilted, vertical edges no longer appear vertical in the image. This feature can however be used to advantage to discriminate amongst straight line edges extracted from the image when searching for landmarks. It might also be used to estimate angles of rotation and distances moved between successive images in order to assist with localisation. Horizontal edges in the real world very rarely appear horizontal in the image due to perspective. By mapping these back to real-world coordinates, it is possible to use the locations of these edges in two successive images to measure rotations or translations of the robot.

Keywords: computer vision, monocular camera, camera tilt, perspective, visual localisation.

5.1 Introduction

The price of digital cameras has plummeted in recent years due to the popularity of web cameras and cameras in mobile phones. This has provided a useful, though complex, sensor for robots.

Unfortunately, these cheap cameras often have poor optics and a limited field of view (FOV) – as little as 40–60°. This is a problem if the robot needs to use the camera to locate obstacles in its path because the camera needs to be tilted downwards. One side-effect is that vertical edges in the real world no longer appear vertical in the camera image.

There are alternatives to using a single camera with a small FOV in order to avoid camera tilt, such as stereo vision, wide-angle lenses or even panoramic or omni-directional cameras. However, these solutions are more expensive and introduce the problems of complex calibration and/or dewarping of the image. Therefore, we persist with a single camera.

T. Taylor: *Applying High-Level Understanding to Visual Localisation for Mapping*, Studies in Computational Intelligence (SCI) **76**, 35–42 (2007)
www.springerlink.com

In the discussion that follows, the term 'vertical' is used to refer to any edge of an object that is vertical in the real world, e.g. a door frame or the corner where two walls meet. This is distinct from a 'horizontal' which refers to an edge that is horizontal in the real world, e.g. the skirting board where a wall meets the floor.

Vertical and horizontal edges do not usually appear as such in a camera image, especially when the camera is tilted. This is a consequence of perspective and is unavoidable.

5.2 Mapping Using Vision

Our objective is for a robot to build a map of an indoor environment using only monocular vision. Several algorithms have emerged in recent years for simultaneous localisation and mapping (SLAM) using vision.

Maps can be built using features observed in the environment. The SIFT (scale-invariant feature transform) algorithm has been used for visual SLAM by several researchers, including the original developers [9]. SIFT features have distinctive signatures making them easy to match. No attempt is made to extract lines from images. In our experience, low-resolution images of empty corridors do not yield many SIFT features.

In [13], the approach is also to identify significant features in images and then determine location by retrieving matching images from a database. To make this robust, they incorporate Monte Carlo Localisation. Although they rely on an algorithm for calculating the signature of selected image features, they also note that features such as lines are useful in general. Clearly this approach requires the image database to be populated in advance and is not so useful for SLAM.

A structure from motion (SFM) approach has also been used. This relies on developing a model of the real world by solving sets of equations based on point correspondences derived from multiple images.

Davison used this approach, and more recently has enhanced his original work by using a wide-angle lens [3]. The key advantage claimed is that the wider field of view allows landmarks to remain in view for longer and therefore contribute to more matches. However, he points out that a special mathematical model is required for the camera because it does not conform to the conventional perspective projection. Furthermore, this approach cannot determine absolute scale so the map stretches as the speed of the camera increases.

In [4] the particle filter approach of FastSLAM is applied to visual SLAM using a single camera. As in the previous methods, the selected features are based solely on how distinctive they are, not on their significance in the real world.

Visual odometry has been demonstrated using a commercial web camera [2]. This relies basically on optic flow. The intention in this case is obstacle avoidance and reliable odometry, but not the repeatable identification of landmarks. Therefore the algorithm does not extract lines or draw a map.

We take a geometric approach to processing lines in images. The selected edges are important because they represent significant features in the environment. This approach follows the seminal work of Faugeras [5] on 3D vision. However, in his work he used matrix notation and dealt with the general case. We have taken specific cases with simple solutions that seem to have gone unnoticed.

It might be tempting to think that a transformation of the image could undo the effects of the camera tilt. However, when a camera moves the images are not related by a projective transformation unless all of the points in the image are coplanar [6]. In other words, it is impossible to correct for perspective across a whole image without knowing the locations of all the objects in the image.

5.3 Vertical Edges

Vertical edges are important as landmarks, especially in man-made environments. These features can be useful for localisation of a robot after small motions.

For a camera that is not tilted, finding vertical edges is trivial because the edges are vertical in the image. However, for a tilted camera this is no longer the case. Furthermore, the slope of these 'vertical' edges in the image is dependent on where the edge is physically located with respect to the camera.

Figure 5.1 shows objects viewed from a tilted camera. Before images can be processed, especially where straight lines are involved, the effects of camera distortion must be removed – the camera must be calibrated, e.g. using [1]. This is particularly important for cameras with cheap lenses which often suffer from some amount of radial distortion.

The left-hand image in Fig. 5.1 is the raw image from the camera, and the right-hand one is after correction for lens distortion. It is clear that the vertical

Fig. 5.1. Variations in slope of vertical edges

edges are not vertical, or parallel to each other. In fact, if all the 'verticals' were extrapolated they would meet at a vanishing point which is well outside the image.

We would like to use the information in the image to assist in identifying the location of obstacles, or the motion of the robot. Therefore, we want to extract vertical edges.

Starting with the standard pin-hole camera equations, e.g. as in [7], it is possible to obtain an equation for the slope of a vertical edge in an image. For the derivation, see [12].

The slope, m, of a line in an image can be calculated using (5.1) below where θ is the camera tilt angle, and $(x,\ z)$ are the co-ordinates of the vertical edge measured across the floor from the camera.

$$m = \frac{-z}{x\ \sin(\theta)} \tag{5.1}$$

Figure 5.2 shows a simulation where the camera has a 60° horizontal FOV and is tilted downwards at 30°. Vertical lines have been drawn from the corners of the checkerboard pattern on the floor. The slopes of these lines change significantly across the image, as expected.

The maximum possible sideways lean of a vertical edge is determined by the FOV of the camera and its height above the floor. This sets a limit on the ratio of z to x for a given camera. For the example above, the 'worst case' slope is around 66°. For instance, most of the verticals in Fig. 5.2 have a slope between 70° and 90°.

By using an edge detector and a line detector, e.g. the Canny edge detector and probabilistic Hough transform in the Intel OpenCV Library [8], it is possible to obtain a set of lines from an image. These can quickly be tested to see if they might represent vertical edges. (Many of the detected lines will not be from verticals.)

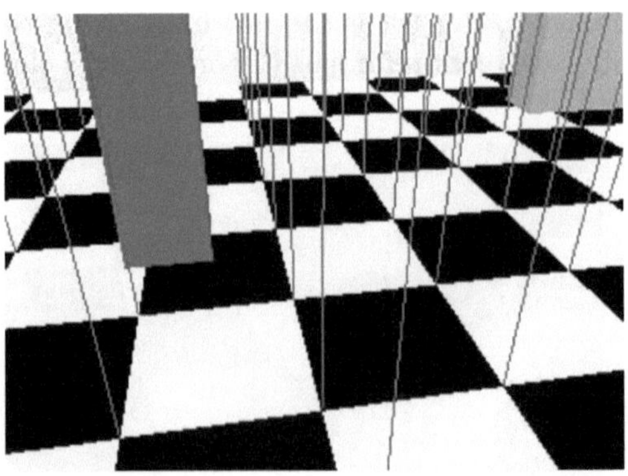

Fig. 5.2. Verticals as viewed by a tilted camera

If the point of intersection of the vertical edge with the floor can be determined accurately, then this can be used to obtain the x and z co-ordinates of the feature using inverse perspective mapping via a simple table lookup [10]. For these coordinates, the slope of the vertical edge can also be calculated (and stored in a lookup table as well), thereby confirming whether or not the intersection point is correct.

Assume that the robot has rotated on the spot, and that the camera is located at the centre of the robot. If a vertical edge is visible in two successive images then the angle of rotation can be expressed simply by the following equation [12]:

$$\psi = \arctan(m_1 \sin(\theta)) - \arctan(m_2 \sin(\theta)) \tag{5.2}$$

To test this, we took photos using a high-resolution camera. A Pentax Optio S7 digital camera was used because it allows for manual focusing – a feature that is not common in digital cameras. (Autofocus would invalidate the assumption of a fixed focal length used in deriving (5.1).)

Two sample images are shown in Fig. 5.3. The resolution of the original images was 7.1 megapixels, or 3072 × 2304.

Using one image as the reference, the rotation angles between it and three other images were calculated. The 'ground truth' angle δ was obtained based on the camera extrinsic parameters. The checkerboard grid provided the necessary reference. Table 5.1 shows the results.

Fig. 5.3. Images obtained by rotating the camera

Table 5.1. Rotation angles calculated from corresponding vertical edges

Test	Angle δ	Angle ψ
1	20.48	20.29
2	5.42	5.15
3	12.74	12.83

As can be seen from the table, the differences between the angles calculated using the calibration grid and those obtained via the formula for vertical edges (in (5.1)) are much less than a degree in all cases. This is much better than the results reported in [12] due to the very high resolution of the camera.

In practice, most of the images in a corridor environment do not contain vertical edges. Also, it is less reliable at lower resolutions.

5.4 Horizontal Edges

Now consider the case where a line in the image results from a horizontal edge in the real world, i.e. the edge between the floor and a wall. These edges are often easy to locate because rooms usually have a ‘kick board’ or cable duct around the edges of the walls. This is normally a different colour from the wall making it easy to distinguish.

Unfortunately, there is no simple formula for the slope of a line in the image corresponding to a horizontal edge. However, by applying inverse perspective mapping [10] the edge can be mapped to real-world co-ordinates.

Figure 5.4 shows the parametric representation of a line in 2D using the orthogonal distance, r, from the line to the origin and the angle, ϕ, between the line and the X-axis.

If the camera is rotated, or equivalently the line is rotated, about the origin, then the orthogonal distance from the origin does not change. The difference in the slopes of the two lines in Fig. 5.4 is the angle of rotation. Therefore, it is possible to calculate the amount of a rotation as the difference between the slopes of the two lines provided that they are first expressed in the real-world robot co-ordinate frame.

On the other hand, if the robot moves forward then no rotation takes place and the two edges should remain parallel.

This allows walls to be used to align the robot by comparing successive images. After a motion, the robot can locate the walls. See [11] for an explanation of how the floor can be detected. The Douglas-Peucker algorithm from

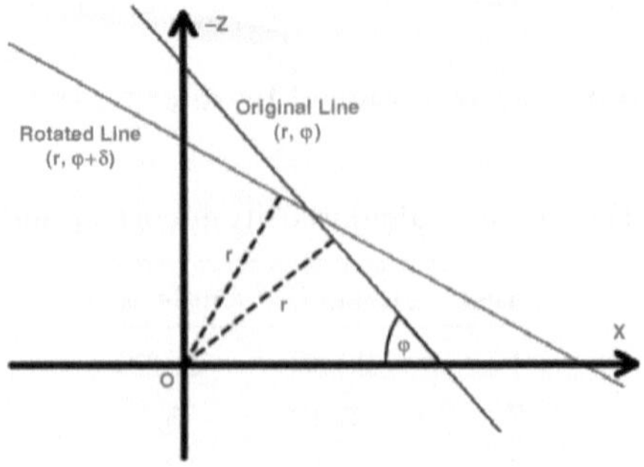

Fig. 5.4. Rotation of a horizontal edge viewed from the top down

OpenCV can be applied to the floor boundary to obtain a set of straight lines. Most of these correspond to walls.

A stronger constraint can be applied in an indoor environment. If it is assumed that the walls are orthogonal in the map, i.e. North–South or East–West, then the robot's orientation can be determined by comparing the estimated angle from a wall with one of these cardinal directions.

Small errors can quickly be corrected, thereby preventing the robot from becoming disoriented. Because errors in rotations have a far greater cumulative effect than errors in position, the ability to obtain accurate orientation estimates is very important.

5.5 Example: Incremental Localisation

Applying the principles outlined above for horizontal edges, Fig. 5.5 shows the improvement in the quality of a map based solely on incremental localisation. Figure 5.5a is based on wheel odometry only.

There is still a problem with the map in Fig. 5.5b because the robot became disoriented in the bottom-right corner. We can correct this using a particle filter, but the explanation is beyond the scope of this chapter.

5.6 Conclusion

This chapter has outlined some simple geometric relationships that can be used to recover vertical edges from an image when the camera is tilted. By comparing the slopes of vertical edges in successive images, it is possible to calculate distances moved or angles of rotation.

Extracting horizontal edges from the floor boundary enables the estimation of rotations. If the assumption is made that walls are orthogonal, then information from horizontal edges can be used to obtain an absolute orientation. This information can be used to assist with localisation as part of a visual SLAM algorithm.

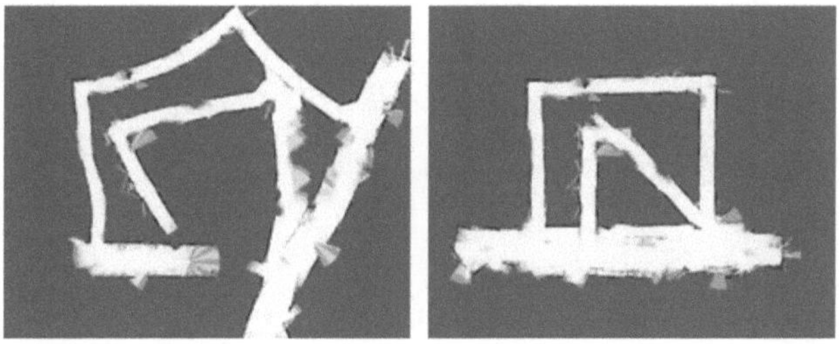

Fig. 5.5. Maps created (**a**) without localization and (**b**) with localisation

References

1. Bouguet J-Y (2006) Camera calibration toolbox for Matlab. http://www.vision.caltech.edu/bouguetj/calib_doc/, visited 19-Mar-2006
2. Campbell J, Sukthankar R, Nourbakhsh I, Pahwa A (2005) A robust visual odometry and precipice detection system using consumer-grade monocular vision. In: Proc IEEE Int Conf on Robotics and Automation, Barcelona, pp 3421–3427
3. Davison AJ, Cid YG, Kita N (2004) Real-time 3D SLAM with wide-angle vision. In: Proc 5th IFAC/EURON Symposium on Intelligent Autonomous Vehicles, Lisbon
4. Eade E, Drummond T (2006) Scalable monocular SLAM. In: Proc IEEE Computer Society Conference on Computer Vision and Pattern Recognition, New York, vol 1, pp 469–476
5. Faugeras O (1993) Three-Dimensional Computer Vision. MIT Press, Cambridge MA
6. Hartley R, Zisserman A (2003) Multiple View Geometry in Computer Vision, 2nd edn. Cambridge University Press, Cambridge UK
7. Horn BKP (1986) Robot Vision. MIT Press, Cambridge MA
8. Intel Corporation (2005) Open Source Computer Vision Library (OpenCV). http://www.intel.com/technology/computing/opencv/, visited 2-Aug-2005
9. Se S, Lowe DG, Little JJ (2005) Vision-based global localization and mapping for mobile robots. IEEE Trans on Robotics, vol 21:3, pp 364–375
10. Taylor T, Geva S, Boles WW (2004) Monocular vision as a range sensor. In: Proc Int Conf on Computational Intelligence for Modelling Control and Automation, Gold Coast, Australia, CD-ROM
11. Taylor T, Geva S, Boles WW (2005) Early Results in Vision-Based Map Building. In: Murase K, Sekiyama K, Kubota N, Naniwa T, Sitte, J (eds) Proc 3rd Int Symp on Autonomous Minirobots for Research and Edutainment, Fukui, Japan, pp 207–216
12. Taylor T, Geva S, Boles WW (2006) Using Camera Tilt to Assist with Localisation. In: Proc 3rd Int Conf on Autonomous Robots and Agents, Palmerston North, New Zealand
13. Wolf J, Burgard W, Burkhardt H (2005) Robust vision-based localization by combining an image retrieval system with Monte Carlo localization. IEEE Trans on Robotics, vol 21:2, pp 208–216

6

Development of an Optical Three-Axis Tactile Sensor for Object Handing Tasks in Humanoid Robot Navigation System

Hanafiah Yussof[1], Masahiro Ohka[1], Hiroaki Kobayashi[2], Jumpei Takata[2], Mitsuhiro Yamano[3], and Yasuo Nasu[3]

[1] Graduate School of Information Science, Nagoya University, Japan
[2] Graduate School of Engineering, Nagoya University, Japan
[3] Faculty of Engineering, Yamagata University, Japan

Summary. Autonomous navigation in walking robots requires that three main tasks be solved: self-localization, obstacle avoidance, and object handling. This report presents a development and application of an optical three-axis tactile sensor mounted on a robotic finger to perform object handling in a humanoid robot navigation system. Previously in this research, we proposed a basic humanoid robot navigation system called the groping locomotion method for a 21-dof humanoid robot, which is capable of defining self-localisation and obstacle avoidance. Recently, with the aim to determining physical properties and events through contact during object handling, we have been developing a novel optical three-axis tactile sensor capable of acquiring normal and shearing force. The tactile sensor system is combined with 3-dof robot finger system where the tactile sensor in mounted on the fingertip. Experiments were conducted using soft, hard, and spherical objects to evaluate the sensors performance. Experimental results reveal that the proposed optical three-axis tactile sensor system is capable of recognizing contact events and has the potential for application to humanoid robot hands for object handling purposes.

Keywords: Humanoid robot navigation, object handling, optical three-axis tactile sensor, optical waveguide.

6.1 Introduction

A humanoid robot is the type of walking robot with an overall appearance based on that of the human body [1]. Humanoid robots are practically suited to coexist with humans because of their anthropomorphism, human friendly design, and locomotion ability. Eventually, the working coexistence of humans and humanoids sharing common workspaces will impose on humanoids with their mechanical control structure the requirement to perform tasks

H. Yussof et al.: *Development of an Optical Three-Axis Tactile Sensor for Object Handing Tasks in Humanoid Robot Navigation System*, Studies in Computational Intelligence (SCI) **76**, 43–51 (2007)
www.springerlink.com

in environments with obstacles [2]. As far as this working coexistence is concerned, a suitable navigation system combining design, sensing elements, path planning, and control embedded in a single integrated system is necessary to guide humanoid robots activities.

Autonomous navigation in walking robots requires that three main tasks be solved: self-localization, obstacle avoidance, and object handling [3]. To realize the robot in the real-world, a sensor-based navigation function is required because the robot is not able to autonomously operate based on environment recognition alone. A tactile sensor system is essential as a sensory device to support the robot control system [4–6]. This tactile sensor is capable of sensing normal force, shearing force, and slippage, thus offering exciting possibilities for determining object shape, texture, hardness, etc. [7]. In this report, we present the development of an optical three-axis tactile sensor system mounted on robotic finger for object handling tasks in contact interaction-based humanoid robot navigation project.

6.2 Contact Interaction-Based Navigation System

It is inevitable that the application of humanoid robots in the same workspace as humans will result in direct physical-contact interaction. This will require the robot to have additional sensory abilities. Besides sensor systems that help the robot to structure their environment, like cameras, radar sensors, etc., a system on the robot surface is needed that enables to detect physical contact with its environment, particularly when the vision sensor is ineffective. In this research, we focus in contact interaction-based navigation applying contact-based sensors with the aim of supporting current visual-based navigation. We have previously proposed a basic contact interaction-based navigation system in humanoid robot called "groping locomotion method" consists of self-localization and obstacle avoidance tasks [8,9]. Six-axis force sensors were attached to both of the humanoid's arms as end-effectors that directly touch objects and provide force data which are subsequently converted to position data by the robot's control system to recognize its surroundings. We used the previously developed 1.25-m tall, 21-dof humanoid robot system named Bonten-Maru II, as experimental platform. Figure 6.1 shows demonstration of the humanoid robot performing self-localization and avoiding obstacle in the groping locomotion method.

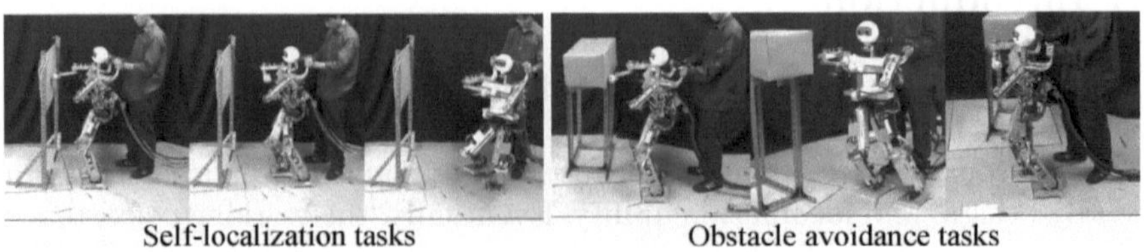

Fig. 6.1. Demonstration of humanoid robot motions in groping locomotion method

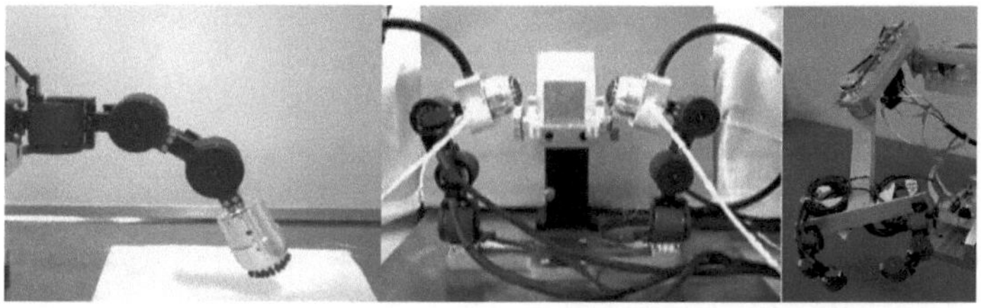

Fig. 6.2. Robotic fingers mounted with optical three axis tactile sensor and application of the finger's system in humanoid robot arm

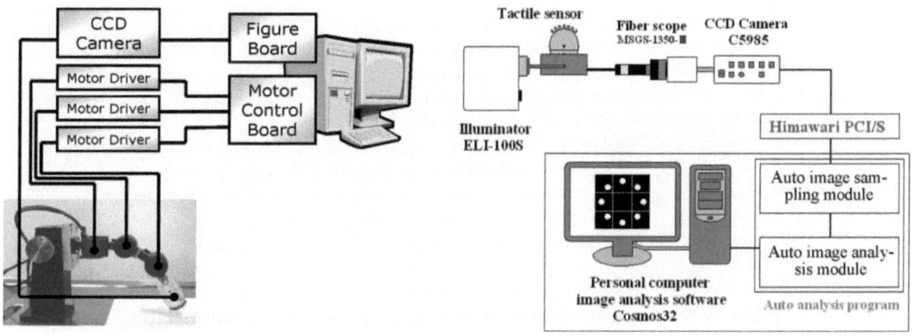

Fig. 6.3. System structure of the 3-dof robotic finger with the optical three-axis tactile sensor and layout of the tactile sensor system

In current research, we focus on development of the object handling tasks. Here we present the development of an optical three-axis tactile sensor capable of acquiring normal and shearing force, with the aim of installing it on a real humanoid robot arm. This report includes an experimental evaluation of the performance of an integrated system comprising the tactile sensor and robotic finger to recognize and grip hard object, soft objects, and also a spherical object. Figure 6.2 shows the robotic fingers mounted with the optical three-axis tactile sensor and the finger's system mounted on humanoid robot arm.

6.3 Optical Three-Axis Tactile Sensor

In this research, we have developed a novel optical three-axis tactile sensor capable of acquiring normal and shearing force with the aim of establishing object handling ability in a humanoid robot navigation system. This sensor is designed to be mounted on a humanoid robot's fingers [10]. We have developed robotic finger comprising of three micro-actuators (YR-KA01-A000, Yasukawa) and connected to a PC via a motor driver and motor control board. The PC is installed with the Windows OS, the image analysis software Cosmos32, and a Visual C++ compiler. Figure 6.3 displays the integrated control

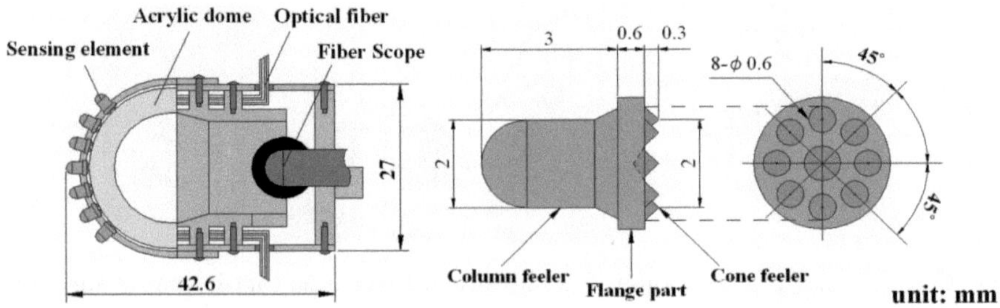

Fig. 6.4. Structure of hemispherical optical three-axis tactile sensor and sensing element

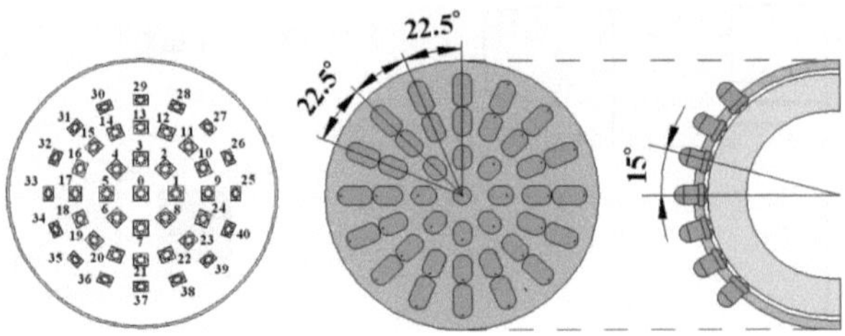

Fig. 6.5. Arrangement of sensing elements on fingertips

system structure of the 3-dof robotic finger and optical three-axis tactile sensor used in the experiment. This figure also displays a layout of the tactile sensor system.

The structure of the optical three-axis tactile sensor used here consists of an acrylic hemispherical dome, an array of silicon rubber sensing elements, a light source, an optical fibre-scope, and a CCD camera, as shown in Fig. 6.4. The silicone rubber sensing element comprises one columnar feeler and eight conical feelers. The eight conical feelers remain in contact with the acrylic surface while the tip of the columnar feeler touches an object. The sensing elements are arranged on the hemispherical acrylic dome in a concentric configuration with 41 subregions as shown in Fig. 6.5, which also displays the structure of the fingertip with the sensing elements. The optical three-axis tactile sensor is based on the principle of an optical waveguide-type tactile sensor. Figure 6.6 shows the sensing principle of the optical three-axis tactile sensor system and image data acquired by the CCD camera. The light emitted from the light source is directed toward the edge of the hemispherical acrylic dome through optical fibers. When an object contacts the columnar feelers, resulting in contact pressure, the feelers collapse. At the points where the conical feelers collapse, light is diffusely reflected out of the reverse surface of the acrylic surface because the rubber has a higher reflective index.

The contact phenomena consisting of bright spots caused by the feelers collapse are observed as image data, which are retrieved by the optical fiber-scope connected to the CCD camera and are transmitted to the computer. The

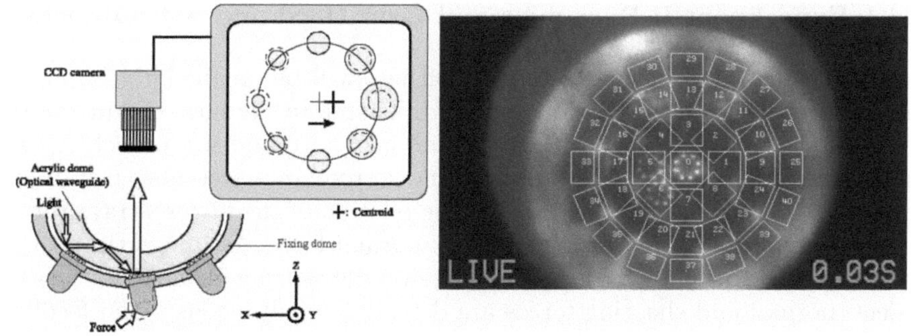

Fig. 6.6. Principle of optical three-axis tactile sensor system and CCD camera-captured image of contact phenomenon in the tactile sensor

dividing procedure, digital filtering, integrated gray-scale value, and centroid displacement are controlled on the PC using auto analysis programme applying the image analysis software Cosmos32. In this situation, the normal force of Fx, Fy, and Fz values are calculated using the integrated gray-scale value G, while shearing force is based on the horizontal centroid displacement. The displacement of the gray-scale distribution u is defined in (6.1), where i and j are orthogonal base vectors of the X and Y axes of a Cartesian coordinate, respectively. Each force component is defined as shown in (6.2).

$$\mathbf{u} = u_x\mathbf{i} + u_y\mathbf{j} \tag{6.1}$$

$$F_x = f(u_x),\ \ F_y = f(u_y),\ \ F_z = g(G) \tag{6.2}$$

6.4 Experiments and Results

Application of tactile sensor in humanoid robot is still under development. Although current robot hands are equipped with force sensors to detect contact force, they do not make use of tactile sensors capable of detecting an object's hardness and/or softness, or can they recognize the shape that they grip. For a robot hand to grip an object without causing damage to it, or otherwise damaging the sensor itself, it is important to employ sensors that can adjust the gripping power.

We conduct experiments to evaluate the performance of the optical three-axis tactile sensor system using the 3-dof robot finger. Unfortunately, the parameters of the human hand and fingers that are involved in sensing the hardness and/or softness of an object have not been fully researched. However, in this experiment we specified the force parameter values to perform force-position control. The minimum force parameter is specified at *0.3* $\sim$ *0.5* *N* (threshold 1), with the maximum side being *1.4* $\sim$ *1.5* *N* (threshold 2). Meanwhile 2 N is fixed as the critical limit. To simplify the analysis, we only consider sensing element number *0* (refer Fig. 6.5).

6.4.1 Experiment I: Recognize and Grip Hard and Soft Objects

The ability to sense hardness and/or softness will be particularly important in future applications of the robotic hand. Therefore, we have conducted this set of experiment in order to recognize and grip hard and soft object. An aluminium block represents a hard object, and styrofoam represents a soft object. Figure 6.7 contains graphs that define the relation of the finger's trajectory at the force applied to the sensing element, normal force, and fingertip position data for each experiment. Here, when the tactile sensing element touches the object, normal and shearing forces are detected simultaneously and feedback to the finger's control system. The finger responds to the applied shearing force

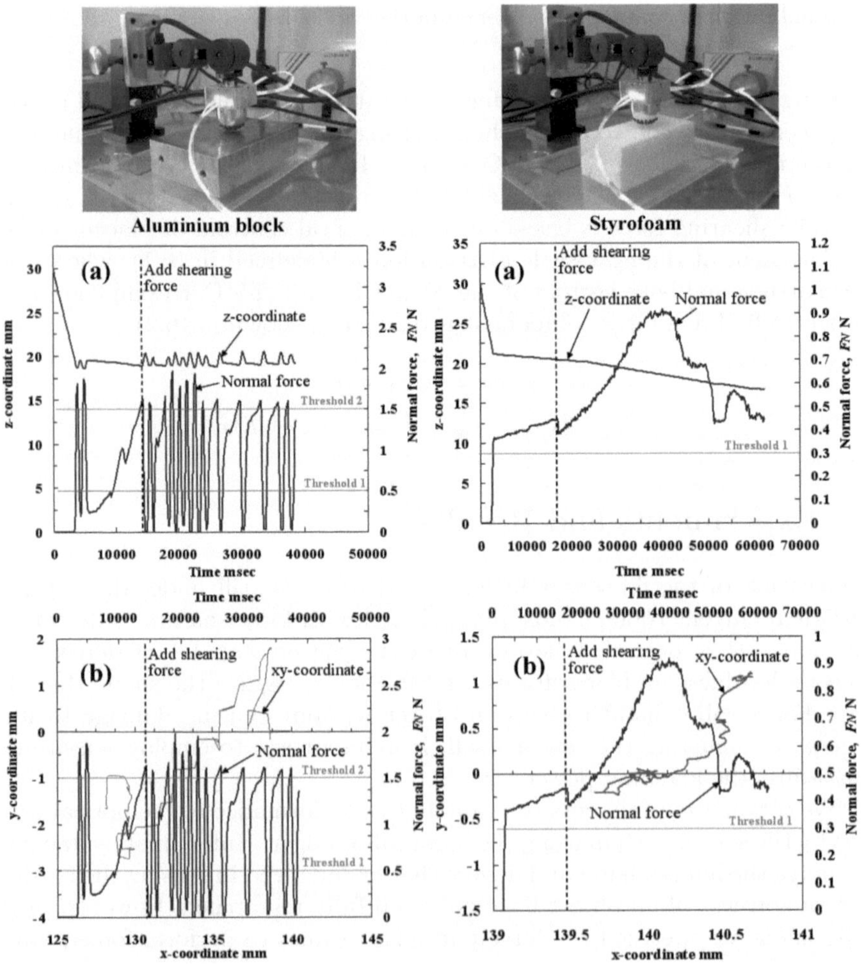

Fig. 6.7. Grip movement of the robot finger on an aluminium block and styrofoam (**a**) Relation between z-coordinate and normal force. (**b**) Relation between xy-coordinate and normal force

by generating trajectory at the XY-axis to define a stable gripping position. Meanwhile, the finger movement at the Z-axis is to adjust the applied normal force to grip the object.

Regarding the aluminium block, some disturbance to the normal force was observed at an early stage as shown in the graph. This phenomenon is the result of the control system adjusting the touching force and the finger is searching for a suitable gripping position. In this situation, due to the hard surface of aluminium block, the force detected by the sensor element is high and exceeds threshold 2. However, the finger feedback system controls the finger movement in order to search for a stable gripping force and position. Finally, the grip force was maintained within the specified force parameter values (between thresholds 1 and 2).

Meanwhile, in the case of styrofoam, because of the surface's softness, the sensor element tended to gnaw into it resulting in a small detected shearing force. Therefore, the movement of finger at the XY-axes for styrofoam is very small as shown in Fig. 6.7b. The sensor element is mainly detecting normal force, where finger movement at the Z-axis can be observed in Fig. 6.7a. However, as the graph indicates, the tactile sensor managed to recognize touch with the soft object where normal force is detected. Finally, the finger movement stopped when the stable grip force was defined. Furthermore, when the stability of the grip force and the position were satisfactory, the finger remained at that position to hold the object. At this point, we purposely added shearing force from the X-axis direction to both objects to define the proposed system's response against slippage. According to the graphs, the finger grip force increased drastically, at which point the tactile sensor detected shearing force. The finger feedback control system responded by adjusting the finger position and the applied force again to define a stable gripping position and force. This phenomenon can be observed in the Fig. 6.7.

6.4.2 Experiment II: Recognize and Grip a Spherical Object

In real-time applications, robots might have to comply with objects of various shapes. In this experiment, we used a tennis ball to evaluate the performance of the proposed tactile sensor and finger system at recognizing a spherical object. In the experiment, the initial touch point was shifted slightly from the peak point of the tennis ball so that the movement characteristic of the robot finger against the sphere's shape surface could be evaluated. Referring to Fig. 6.8, when the touch force increased, the rotation moment acted on the tennis ball, causing the ball to try to rotate. The shearing force increased simultaneously and when the applied normal force exceeded the threshold 2, the ball started to slip out. Hence, the finger corrected its position to move nearer to the object's surface and adjust the gripping force. This process was repeated, finally resulting in a change to the finger orientation in order to comply with the spherical surface shape to stop when the finger reached a stable gripping position.

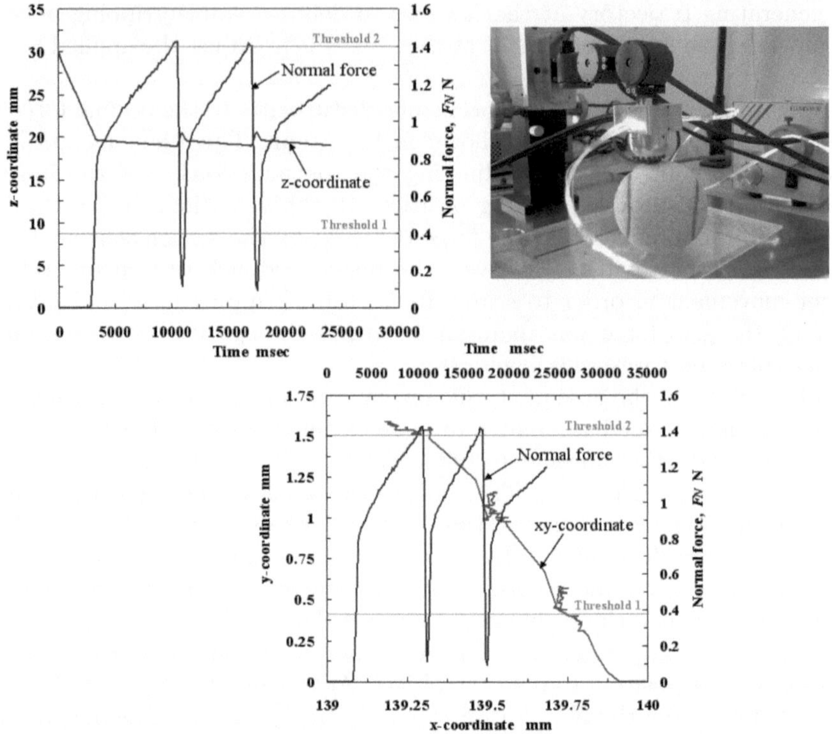

Fig. 6.8. Grip movement of the robot finger on tennis ball

Fig. 6.9. Grip and rolling movement of the robot finger on real egg

In addition to above, we conducted an experiment to evaluate the force control performance of the integrated system on a spherical and fragile object. We used a real egg as shown in Fig. 6.9. Experimental result reveals good performance of the force feedback control to grip and roll the egg without breaking it.

6.5 Conclusion

The optical three-axis tactile sensor system presented in this research is capable of clearly acquiring normal force and shearing force. Furthermore, it also can detect force at the *XYZ*-axes of a Cartesian coordinate. These characteristics have given the proposed tactile sensor the possibility of recognizing the hardness and/or softness of an object, as presented in Sect. 6.4.1. Next, according to experimental results in Sect. 6.4.2, the proposed system is also capable of recognizing spherical shape and then adjusts the finger's gripping position and force it to hold the object. In addition, experiment to grip and manipulate a fragile object revealed good performance of the system's force feedback control. Our overall conclusion from both experimental results revealed that the optical three-axis tactile sensor system, combined with the newly developed 3-dof robot finger, can be readily applied to a robot's hand, and that the robot will be capable of controlling its gripping power by using the feedback algorithm of this embedded system.

Future work will involve further development of the contact interaction-based navigation project, by applying the integrated system comprising the optical three-axis tactile sensor and two robot fingers to perform object handling tasks. It is anticipated that using this novel tactile sensor system technology will bring forward the evolution of human and humanoid robots working together in real life.

References

1. Hirai K, Hirose M, Haikawa Y, Takenaka T (1998) The development of Honda hu-manoid robot. Proceedings of ICRA'98: 1321-1326
2. Vukobratovic M, Borovac B, Babkovic K (2005) Contribution to the study of anthropomorphism of humanoid robots. Journal of Humanoids Robotics 2(3): 361–387
3. Clerentin A (2005) Self localization: a new uncertainty propagation architecture. Journal of Robotics and Autonomous Systems 51(2–3): 151–166
4. Hanafiah Y, Ohka M, Kobayashi H, Takata J, Yamano M, Nasu Y (2006) Contribution to the development of contact interaction-based humanoid robot navigation system: Application of an optical three-axis tactile sensor. Proc. 3rd ICARA06: 63–68
5. Omata S, Murayama Y, Constantinou CE (2004) Real time robotic tactile sensor system for determination of the physical properties of biomaterials. Journal of Sensors and Actuators A, 112(2–3): 278–285
6. Kerpa O, Weiss K, Worn H (2003) Development of a flexible tactile sensor system for a humanoid robot. Proceedings IROS2003, vol. 1: 1–6
7. Lee MH, Nicholls HR (1999) Tactile sensing for mechatronics: a state of the art survey. Journal Mechatronics 9(1): 1–31
8. Hanafiah Y, Yamano M, Nasu Y, Ohka M (2005) Obstacle avoidance in groping lo-comotion of a humanoid robot. Journal of Advanced Robotic Systems, 2(3): 251-258
9. Hanafiah Y, Yamano M, Nasu Y, Ohka M (2006) Humanoid robot navigation based on groping locomotion algorithm to avoid an obstacle. Mobile Robots, Towards New application, ARS International and pro literature Verlag, Germany
10. Ohka M, Kobayashi H, Mitsuya Y (2006) Sensing precision of an optical three-axis tactile sensor for a robotic finger. Proceeding 15th RO-MAN2006: 220–225

7

Clustering Methods in Flock Traffic Navigation Based on Negotiation

Carlos Astengo-Noguez and Ramón Brena-Pinero

Center of Artificial Intelligence
ITESM, Campus Monterrey, México
{castengo,ramon.brena}@itesm.mx

Summary. Everyday people face traffic congestion in urban areas, raising travel time and stress issues in drivers. Beyond traffic lights synchronization, emerging location-based technologies like GPS and cellular communications suggest some futuristic traffic coordination schemes. Flock traffic navigation (FTN) based on negotiation gives a basic algorithm and showed by simulation that a non-centred solution can be achieved letting individual vehicles to communicate and negotiate among them. Early works suppose a bone diagram formed by two agents and their geometrical intersection based on their initial and end points. In this research paper we are proposing new methods based on clustering in order to allow the entrance of new agents into the bone structure and flock fragmentation if it is convenient.

Keywords: Navigation, Flock traffic, GPS, Clustering method, Negotiation, K-means algorithm.

7.1 Introduction

Everyday people face traffic congestions in big cities. Currently, traffic is controlled through traffic-lights, static signs and, in a few places, electronic boards that give information about traffic flow, road accidents or weather-related situations. Traffic congestion describes a condition in which vehicle speeds are reduced below normal, increasing drive times, and causing vehicle queuing. It occurs only when the demand is greater than the roadway's capacity [14].

Over the last 30 years, the number of cars per 1 000 persons has doubled and the distance travelled by road vehicles has tripled. Europe estimates 300 million drivers and daily 10% of its roads are congested [1].

Reducing total congestion saves time and fuel, and leads to decreased vehicle emissions [6]. The Texas Transportation Institute estimates that in 75 of the largest US cities in 2001, $69.5 billion dollars are wasted in time and fuel costs [12].

C. Astengo-Noguez and R. Brena-Pinero: *Clustering Methods in Flock Traffic Navigation Based on Negotiation*, Studies in Computational Intelligence (SCI) **76**, 53–60 (2007)
www.springerlink.com

Emerging location-based technologies, using the global positioning system (GPS) and the cellular communications technologies allows us to envision some futuristic coordination-based traffic handling mechanisms that could be feasible with current research-level technology [4, 15].

Some agent-based automated vehicle coordination mechanisms have been proposed [5] but they lead to intricate traffic merging. Indeed, in high density traffic, alternating individual cars at intersections could produce very dangerous situations; even assuming the protocol itself is flawless, any factor outside the protocol, like a blown tire or mechanical failure, would be a recipe for disaster. Human lives could not be exposed to such a risk.

Flock traffic navigation (FTN) based on negotiation vehicles could navigate automatically in groups called "flocks" [2]. Birds, animals, and fish seem to organize themselves effortlessly by travelling in flocks, herds, and schools that move in a very coordinated way [13]. A flock consists of a group of discrete objects moving together. In the case of bird flocks, evidence indicates that there is no centralized coordination, but a distributed control mechanism.

FTN allows the coordination of intersections at the flock level, instead of at the individual vehicle level, making it simpler and far safer. To handle flock formation, coordination mechanisms are issued from multiagent systems.

7.2 Flock Traffic Navigation

Birds coordinate their migration behavior in flocks. A similar strategy has been used in [2] for vehicular traffic modelling.

In the city context, flock navigation means that vehicles join together in groups (flocks) to travel through intersections. But as individual cars can have different departure and destination points, we can see that they would travel together for only a part of their entire travel.

Flock navigation, could seem to require automated driving cars, but actually flock navigation could just as well be suggested to human drivers, who would decide whether to follow the recommendation or not.

FTN based on negotiation starts with quite simple idea: people in their vehicles share the same city (or at least a part of it), and if they agree to cooperate with those who have common goals ("near" destination points) then they can travel together forming flocks. Depending of the flock size they will be benefited with a speed bonus (social bonus).This bonus is known a priori.

The mechanism to negotiate [2] starts with an agent who wants to minimize its journey time. He/she knows its actual position and its goal (final position). Because city topology is known he/she can estimate his journey time. Then, he/she broadcast and shares this information in order to look for partners. Neighbor-agents receive and analyse this message. The environment (city) gives a social bonus in terms of maximum speed allowed if agents agree to travel together.

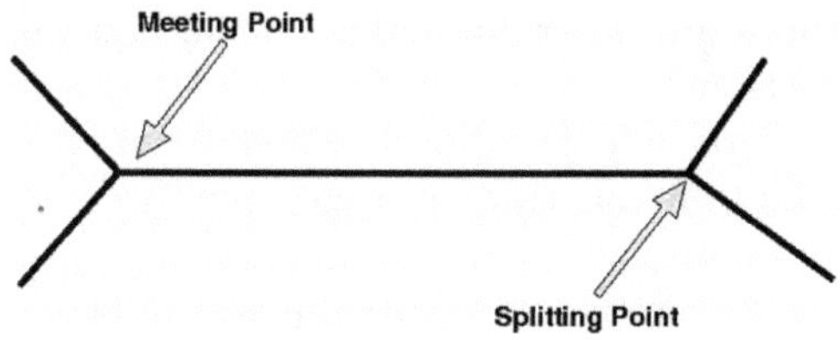

Fig. 7.1. Bone-structure diagram

Individual reasoning play the main role of this approach. Each agent must compare its a priori travel time estimation vs. the new travel time estimation based on the bone-structure diagram (Fig. 7.1) and the social bonus and then make a decision.

Decision will be taken according if agents are in Nash equilibrium (for two of them there is no incentive to choose other neighbor agent than the agreed one) or if they are in a Pareto Set (a movement from one allocation to another can make at least one individual better off, without making any other individual worse off) [16].

If both agents are in Nash equilibrium, they can travel together as partners and can be benefited with the social bonus. In this moment a new virtual agent is created in order to negotiate with future candidates as in [10]. Agents in a Pareto Set can be added to this "bone-structure" if its addition benefits him without making any of the original partners worse off.

Simulations indicate that flock navigation of autonomous vehicles could substantially save time to users and let traffic flow faster [2].

With this approach new agents in the Pareto set who want to be added into the bone-structure must share the original splitting point calculated with the first two agents who were in Nash equilibrium. But new problems can arise.

Could agents in a Pareto set renegotiate within flock members in order to improve their individual benefit having more than a single splitting point? (Flock fragmentation)

Could other agents not considered in the original Pareto set be incorporated as long at the flock travels using rational negotiation with the virtual agent?

We intend to solve these problems improving the mechanism proposed in [2] in the context of rational decision making.

Rational perspective has some advantages over a physics particle perspective where vehicles would simply obey laws regardless of whether or not their behavior agrees with their interests. In a human society, if drivers have the option of driving alone or in automated flocks, they will only do the latter if it agrees with their interests. So, we have to design society mechanisms in such a way that they are individually rational.

As in [2], we will assume a city with the following rules:

- As for birds, there are no traffic lights (or an equivalent control mechanism) at intersections
- Social rules give larger flocks priority when arriving to an intersection
- As a result of the previous rule, average speed will be a function of flock size
- A protocol for intersection control is supposed to be like the one described in [5], but alternating flocks instead of individual vehicles

7.3 Splitting Point Extension

If several agents can be added to the bone-structure then, there could be subnegotiations among the agents and several problems must be answered during the bone-structure trip: Can be more than one splitting point? Where will be localized? Fragmentation can be allowed? The reservation-based model still works for the flock?

7.3.1 Clustering Techniques

Clustering are mathematical techniques that partitioning a data set into subsets (clusters). It is an unsupervised classification of patterns (observations, data items, or feature vectors) into groups or clusters [3, 7].

Cluster analysis is the organization of a collection of pattern (usually represented as a vector of measurements or a point in a multidimensional space) into clusters based on similarity. It is an exploratory data analysis tool which aims at sorting different objects into groups in a way that the degree of association between two objects is maximal if they belong to the same group and minimal otherwise [9].

The data in each subset share some common proximity according to some defined distance measure.

A number of methods and algorithms have been proposed for grouping multivariate data into clusters of sampling units. The methods are exploratory and descriptive, and their statistical properties do not appear to have been developed [9].

Typically, cluster methods are highly dependent on the sampling variation and measurement error in the observations. Small perturbations in the data might lead to very different clusters. The choice of the number of clusters may have to be made subjectively and may not follow from the algorithm.

The clustering process key is the measures of the distances of the observation vectors from one another. If we begin with N objects (agents) the distance may be summarized in the N×N symmetric matrix

$$D = \begin{bmatrix} 0 & d_{12} & \dots & d_{1N} \\ d_{12} & 0 & \dots & d_{2N} \\ \dots & \dots & \dots & \dots \\ d_{1N} & d_{2N} & \dots & d_{NN} \end{bmatrix} \tag{7.1}$$

There are several clustering methods described in the literature as in [8,9, 17]. A taxonomy of these methods are presented in [7]. We will describe those which we believe have a close relation with our problem.

7.3.2 Single-Linkage Algorithm

It combines the original N single-point clusters hierarchically into one cluster of N points. Scanning the matrix for smallest distances and grouping the observations into clusters on that basis. Single-linkage algorithm is a hierarchical clustering method [17].

Input: N travel partners according to [2] negotiation algorithm.

1. For each pair of agents a_i and a_j calculate their distance $d_{i,j}$ using the Manhattan metric:

$$d_{ij} = ||x_i - x_j|| + ||y_i - y_j|| \tag{7.2}$$

and form the D matrix as in (1).

2. Pairing the two agents with smallest distance.
3. Eliminate rows and columns in D corresponding to agents a_i and a_j.
4. Add a row and column of distances for the new cluster according to:
5. For each observation k (agent a_k)

$$d_{k(i,j)} = \min(d_{ki}, d_{kj}),\ k \neq i, j \tag{7.3}$$

The results is a new $(N-1) \times (N-1)$ matrix of distances D. The algorithm continues in this manner until all agents have been grouped into a hierarchy of clusters. The hierarchy can be represented by a plot called a dendrogram.

7.3.3 K-Means Algorithm

It consists of comparing the distances of each observation from the mean vector of each of K proposed clusters in the sample of N observations. The observation is assigned to the cluster with nearest mean vector. The distances are recomputed, and reassignments are made as necessary. This process continues until all observations are in clusters with minimum distances to their means.

The typical criterion function used in partitional clustering techniques is the squared error criteria [15]. The square error for clustering can be defined as

$$e^2 = \sum_{j=1}^{K} \sum_{i=1}^{Nj} ||x_i^{(j)} - c_j||^2 \tag{7.4}$$

where $x_i^{(j)}$ is the i-th agent belonging to the j-th cluster and c_j is the centroid of the j-th cluster. K-means algorithm is an optimization-based clustering method [17].

Input: K number of clusters and clusters centroids.

1) For each new agents a_i assign it to the closest cluster centroid.
2) Re-compute the cluster using the current cluster memberships.
3) If a convergence criterion is not met go to step 2. Typical convergence criteria are minimal reassignment of agents to new cluster centers or minimal decrease in squared error.

Variations of the algorithm were reported in the literature [7]. Some of them allow criteria splitting and merging of the resulting clusters. Cluster can be split when its variance is above a prespecified threshold and two clusters are merged when their centroids is below another prespecified threshold.

7.4 New Flock Traffic Navigation Algorithm Based on Negotiation

Now we can propose a new algorithm based on [2] and the previous clustering algorithms in Sects. 7.3.2 and 7.3.3.

Suppose that in time to there are N agents in a neighborhood and each agent make a broadcast searching for partners according to the distance radius and the algorithm described in [2, 11].

Suppose that agent a_i and agent a_j are in Nash equilibrium. Because individually they are making rational decisions, they agree to travel together using the bone-diagram algorithm which includes a meeting and splitting point for them and the creation of a virtual agent called the flock spirit.

New agent members (which are in the Pareto set) can be added to the flock if new negotiations between each agent and the flock spirit are made in a rational basis.

During the travel time, new negotiations can be done using the original ending-points for each agent for a dynamical clustering procedure using the single-linkage algorithm explained in Sect. 7.3.2 allowing flock fragmentation and so generating new k flock spirits.

New agent-partners which not participle in the earliest negotiations (so, they are not in the Pareto set) can negotiate with the flock spirit using the centroids of the k clusters according to the K-means algorithm described in Sect. 7.3.3.

Input: A City graph, L agents in a neighborhood each one with a start and goal point.

Part A, at t=to

1) Agents broadcast their positions and ending points (goals) and using the bone-diagram algorithm as in [2] find partners in Nash equilibrium.

2) Create the bone diagram, a meeting point and a splitting point. Also create a flock spirit.

Part B, at $t = to + \Delta t$

3) For the rest of the agents that are in a Pareto Set use single-linkage algorithm in order to create new splitting points for each of the k-clusters.
4) Within the flock renegotiate in order to create new k-flock spirits.
5) Calculate the centroid for each cluster.

Part C, at $t > to + \Delta t$

6) For new agents within the broadcast of the flock spirit use K-means clustering algorithm for negotiation basis.

7.5 Concluding Remarks and Future Work

This chapter presents an extended algorithm for flock traffic negotiation that allows new vehicles to enter in a preestablished negotiation.

Using clustering algorithms allow new coming vehicles share the social bonus and give new chances for reallocate splitting points allowing flock fragmentation.

Because individually rational negotiation was done according to [2], re-negotiation within flock members only can improve their individual score.

Using the single-linkage clustering algorithm it is clear that new vehicles (agents) incorporated from the Pareto set actually could change the original agreed splitting point leaving K new splitting points localized at the clusters' centroids.

The K-Means clustering algorithm can be used for agents not considered in the original Pareto in order to determine if individually they can use the flock bone structure for its personal benefit.

With this new algorithm which include the two clustering algorithms the bone structure is no longer a static path with only one splitting point because fragmentation in not only allowed but also desirable from the individual rational point of view.

The Dresner and Stone reservation algorithm [5] still work under these considerations because at intersections there were not any changes to the algorithm assumptions.

Future work includes extending cluster distances outputs (dendrograms distances) variances and square error criteria as probability measures for flock survival; so new agents can consider fragmentation probability for negotiation.

References

1. A. Amditis, "Research on Cooperative Systems-Overview of the on-going EU activities". IEEE- ITSS Newsletter Vol. 8, No. 3, pp 34–37 (September 2006).
2. C. Astengo and R. Brena "Flock Traffic Navigation based on Negotiation" ICARA 2006, Third International Conference on Autonomous Robotics and Agents (ICARA), pp. 381–384, (2006).
3. P. Berkhin, "Survey of Clustering Data Mining Techniques". Technical report, Ac-crue Software, pp 1–56, (2002).
4. S. Biswas et al, "Vehicle-to-vehicle wireless communication protocol for enhancing highway traffic safety", IEEE Communications Magazine, 44(1), pp 74–82 (2006).
5. K. Dresner and P. Stone, "Multiagent traffic management: an improved intersection control mechanism". Proceedings 4th International Joint Conference on Autonomous Agents and Multiagent Systems, Utrecht, ACM Press, New York, pp 471–477 (2005).
6. Federal Highway Administration, "Traffic congestion and reliability, linking solutions to problems", final report 2003, http://www.ops.fhwa.dot.gov/ congestion_report_04/, visited on 1/3/2005 .
7. AK. Jain et al, "Data Clustering: A Review". ACM Computing Surveys, Vol 31, No 3, pp. 265–323. (September 1999).
8. P. Lewicki and T. Hill. "Statistics: Methods and Applications". First edition, Statsoft (2006) pp. 115–126.
9. DF. Morrison "Multivariate Statistical Methods"Data Clustering: A Review". Third edition, McGraw Hill, pp. 385–402. (1990).
10. R. Olfati-Saber, "Flocking for multi-agent dynamic systems: algorithms and theory". IEEE Transactions on Automatic Control, 51(3), pp 401–420, (2006).
11. GVN. Powell, "Structure and dynamics of inter- specific flocks in a neotropical mid-elevation forest", The Auk, 96(2) pp 375–390 (1979).
12. D. Schrank and T. Lomax. "Annual urban mobility report", http://www.pittsburghregion.org/public/cfm/library/reports/2003UrbanMobilityStudy.pdf, visited on 15/4/2005.
13. CE. Taylor and DR. Jefferson. "Artificial life as a tool for biological inquiry", Artificial Life,volume 1(1–2), pp 1–14, (1994).
14. TD. Wang, and C. Fyfe, "Traffic jams: an evolutionary investigation," IEEE/WIC/ACM International Conference on Intelligent Agent Technology (IAT'04), Beijing, pp 381–384 (2004).
15. L. Wischholf et al, "Self-organizing traffic information system based on car-to-car communication: 1st International Workshop on Intelligent Transportation (WIT 2004), Hamburg, pp 49–53 (2004).
16. M. Wooldridge. Introduction to MultiAgent Systems. John Wiley and Sons, pp 129–148. (2002).
17. AM.Yip et al, "Dynamic Cluster Formation Using Level Set Meth-ods". IEEE Transactions on pattern analysis and machine intelligence, Vol 28, No. 6. pp. 877–889 (June 2006).

8

Using Learned Features from 3D Data for Robot Navigation

Michael Happold[1] and Mark Ollis[2]

[1] Applied Perception, Inc., Cranberry Township, Pennsylvania 16066, USA
happ@appliedperception.com

[2] Applied Perception, Inc., Cranberry Township, Pennsylvania 16066, USA
mollis@appliedperception.com

Summary. We describe a novel method for classifying terrain in unstructured, natural environments for the purpose of aiding mobile robot navigation. This method operates on range data provided by stereo without the traditional preliminary extraction of geometric features such as height and slope, replacing these measurements with 2D histograms representing the shape and permeability of objects within a local region. A convolutional neural network is trained to categorize the histogram samples according to the traversability of the terrain they represent for a small mobile robot. In live and offline testing in a wide variety of environments, it demonstrates state-of-the-art performance.

Keywords: Autonomous Systems, Vision Systems for Robotics, Artificial Neural Networks.

8.1 The Challenge for Robot Navigation

The complexity and irregularity of natural environments present a formidable challenge to robot vision systems. This is particularly true because robot vision systems must operate in real-time in order to be of any use for practical applications. Therefore, whether using imagery or range data, it is important to be able to reduce the complexity of the scene to a small set of compact representations. A common method of extracting geometric features from range data is to separate points belonging to an approximate ground plane in the vicinity of the robot from points possibly belonging to obstacles [1]. Once these two classes of points are determined, features such as the maximum height and the slope within a local region can be easily and rapidly computed. Often, these two measurements alone are employed by hand-coding appropriate thresholds that encapsulate the robot's capabilities. This methodology, however, tends to be quite sensitive to noise, especially with respect to the ground plane estimate. And it is almost a certainty that

M. Happold and M. Ollis: *Using Learned Features from 3D Data for Robot Navigation*, Studies in Computational Intelligence (SCI) **76**, 61–69 (2007)
www.springerlink.com

there will be noise in this estimate. Even given a reliable ground plane estimate, it is rather uncertain that such a limited set of geometric measurements can accurately represent the diversity of natural terrain types and their traversability in relation to a particular mobile robot. The tendency has been to ignore this diversity and force the environment into a binary representation. Depending on how one sets the thresholds on the feature measurements, this approach tends to produce either overly-aggressive or overly-conservative behaviour. What is desirable, rather, is to keep the robot safe while exploiting the full range of its capabilities.

In our work, we apply a convolutional neural network to autonomously learn which features are relevant to the task. However, rather than using raw imagery as input as in [2], our classifier operates on 2D histograms of 3D points within local regions as well as 2D histograms of ray pass-throughs associated with points outside of these regions (measuring the permeability of a region). The ground plane estimate remains, though it is used only for approximate placement of the window governing the histogram extraction. The 2D histograms are extracted from a database of stereo range data collected with a small mobile robot in diverse environments. The network learns to mimic human classification of the terrain in labelled regions, extracting its own features from the histograms. It is able to learn complex classifications that easily thwart systems using simple geometric features, and is robust to inaccurate ground plane measurements due to its shift invariance. Because the system is operating on 3D data, placing our terrain classifications into a map becomes trivial. The trained network has been incorporated into the robot's terrain classification system, where it runs in real-time and demonstrates performance superior to classifiers that rely on geometric features such as height and slope.

8.2 Approach

Our system operates in four stages. First, 2D histograms of 3D points and pass-throughs are extracted from evenly spaced regions in the point cloud provided from the stereovision system. These histograms are then range-compensated to account for the decrease in point density as range increases. They are then fed to a convolutional neural network trained to classify the histograms into one of four traversability classes. Finally, the output of the network is converted to a traversability cost and placed in a map used by the robot's planning system.

8.2.1 2D Histogram Extraction

We have chosen 2D histograms because they provide a compact representation of the 3D structure and appearance of regions in a scene. They are accumulated from the projections of 3D points onto 2D slices through the range data, or from the intersections of rays passing through a 2D window to points beyond the window's defined local neighbourhood. Point projection histograms have been used to pick out symmetric structures such as concertina wire from vegetation by comparing the histograms of points on either side of a 2D slice using normalised correlation [3]. We are unaware of any prior use of pass-through histograms.

We use an occupancy grid to partition the range data into columns of 20 cm width $\times$ 20 cm length and infinite height. A 2D histogram is computed by first positioning its horizontal centre at the horizontal centre point of a column and setting its vertical position to be such that the ground plane intersects the window at the column's centre 20 cm above the window's lowest point. The window is divided into 14$\times$80 (width $\times$ height) cells of size 2$\times$2 cm^2, and thus represents a 0.28$\times$1.6 m^2 planar region. A window centred on one column will thus overlap slightly with neighbouring columns.

A human expert hand-labels selected columns in the training data with one of four cost class labels representing increasing cost of traversal, with the highest cost understood to represent untraversable terrain. Three histograms are extracted for each selected region in the training data. One of these derives from a 2D window (PW), shown in Fig. 8.1, that is parallel to the vector from the reference stereo camera to the region centre and the world Z (up) vector and whose centre is aligned with the centre of the region. Points that lie within 14 cm on either side of this window are projected onto their corresponding cell, creating a histogram of the point count. A second histogram is created from a 2D window (OW), also shown in Fig. 8.1, that is orthogonal both to the first plane and the horizontal vector from stereo reference camera to a point directly above the region centre, i.e. having the same Z value as the reference camera. Again, points within 14 cm on either side of the window are projected onto its cells and tallied. The final histogram (RW), shown in Fig. 8.2, represents the counts of the rays from the reference camera to 3D points that pass through this second window beyond its 14 cm inclusion region.

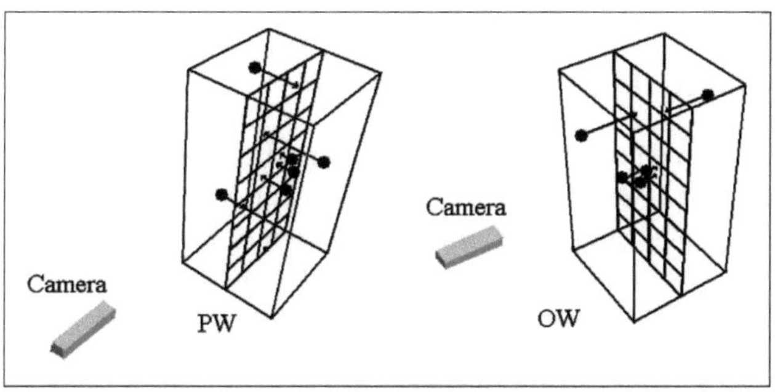

Fig. 8.1. Point projection onto PW (*left*) and OW (*right*) histograms

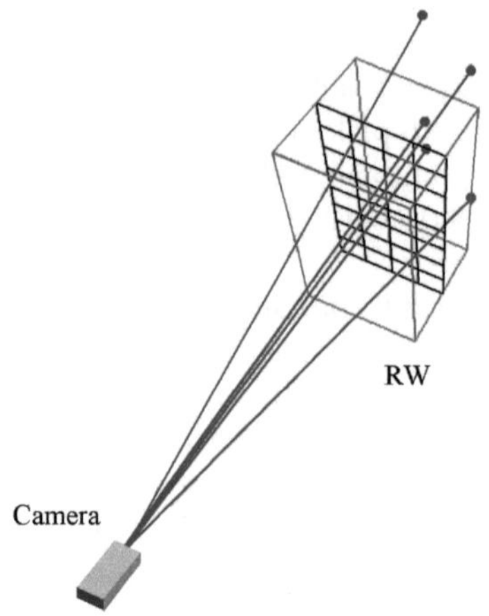

Fig. 8.2. Ray intersection for RW histogram

8.2.2 Range Invariance

Because pixel rays diverge, fewer points will fall within the histograms' local neighbourhoods as range increases. Growing the size of the histogram windows and their respective cells as range increases would account for the sparseness of range data at greater distances, but would in turn introduce a scale variability as objects would cover smaller regions of the windows as they grow. Instead, we account for range variability by scaling the point and ray intersection counts within cells by a function of range. The area of a 2D region subtending the stereo field of view is directly proportional to the square of the range. The number of pixel rays intersecting a 2D window of fixed size is therefore indirectly proportional to the square of its range from the stereo camera. This suggests that we ought to adjust histogram counts by the square of the range R. In order to test the efficacy of this method, we compare classification after adjusting histogram counts by a factor of R, $R \times \sqrt{R}$ and R^2.

8.2.3 Convolutional Neural Network

Convolutional neural networks were developed to relieve designers of classification systems of the need to determine an optimal set of feature extractors as well as to handle the scale and shift invariance of inputs such as handwritten characters [4]. What distinguishes a convolutional neural network is the introduction of convolutional and subsampling layers. Convolutional layers consist of feature maps, each with its own set of weights shared among units within a map. The shared weights act as a trainable convolutional mask, and each unit within a feature map computes the response of a local receptive field within the input to the mask. A trainable bias, shared by all units within a map, is added to the response and a nonlinearity is applied. Because each unit within a map applies the same mask, the maps can be viewed as feature detectors that are invariant to position within the input. Maps in convolutional layers are often connected to multiple input maps, and weight sharing may occur across these multiple connections as well.

Further invariance to scale is introduced by the subsampling layers, which typically follow convolutional layers. Again, the subsampling layers consist of maps with a shared weight, and units within a map are connected to local receptive fields in the previous layer. However, unlike with convolutional layers, these fields do not overlap, so the subsample layer downsamples the previous layer. Each map in a subsample layer is typically connected to a single input map from the previous layer.

We trained six-layer convolutional neural networks on various combinations of these histograms, with the most basic taking the OW point count histogram as input. The convolutional layers learn shift invariant feature extractors, which removes the need to guess which features should be derived from the histograms. The shift invariant nature of the feature extraction is also essential in increasing the system's robustness to ground plane estimates. Training is performed with a variant of RPROP [5] suitable for convolutional networks. RROP uses an adaptive, weight-specific update value for changing weights that eliminates the possibly harmful effects of the magnitude of the partial derivative $\frac{\partial E}{\partial w_{ij}}$. Only the sign of the partial derivative is considered.

The architecture for the simplest network is illustrated in Fig. 8.3, showing a single map for each layer and a half-sized input. In many ways, it mimics that of LeNet5 [4], which is used for character recognition. The full input to our network is a 14×80 histogram. The first hidden layer C1 consists of six convolutional maps, each with its own set of 5×5 weights shared among units within a map. The next hidden layer S1 consists of six subsample maps, one for

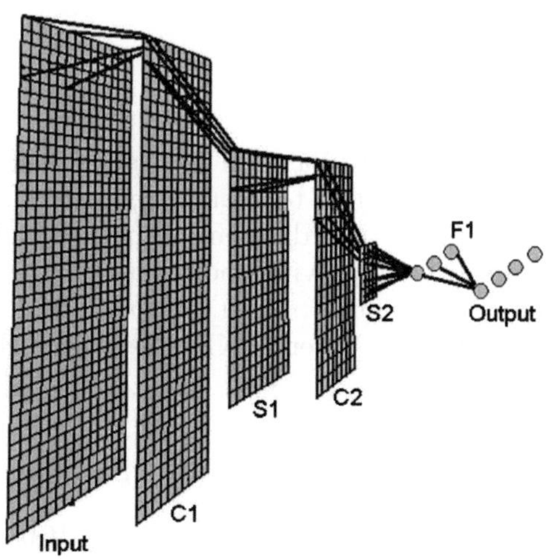

Fig. 8.3. Portion of the OW architecture showing half input size (14×40)

each convolutional map in the previous layer. Each unit in a subsample map is connected to a unique 1×2 set of units in its corresponding convolutional map. The third hidden layer C2 consists of 16 convolutional maps, each connected to a unique subset of the previous layer's maps with which it convolves a 3×3 set of weights unique to each convolutional map but shared across connections to the subsample maps. The fourth hidden layer S2 consists of 16 subsample maps, each connected to a single map in the previous layer and subsampling 2×4 regions. The fifth hidden layer F1, containing 100 units, is fully connected to every unit of every map in the fourth hidden layer. Finally, the output layer is fully connected to the fifth hidden layer. Two more complex networks use either the PW or RW histograms as additional input to a larger network.

8.2.4 Cost Conversion and Terrain Maps

The outputs of the neural network represent cost classes, which in turn are mapped to costs relevant to a D* planner. We convert the raw network output to terrain cost either by using a winner-take-all scheme or by taking the sum of the cost classes weighted by the normalised network output. Costs are then placed in a map consisting of $20\times20\,\text{cm}^2$ cells. As new costs arrive for a cell, a decision must be made how to fuse them with existing data from earlier time steps or from a different eye. We have developed a confidence metric based on position within the stereo FOV to determine which cost update is kept in a cell. This offers a degree of perceptual stability that alleviates one of the practical drawbacks of replanning algorithms such as D* (which we run on our map to determine a global path), namely constantly changing planned paths in response to new, conflicting updates.

8.3 Results

8.3.1 Platform

The data collection and testing platform is a small outdoor mobile robot developed for the Learning applied to ground robotics (LAGR) program by Carnegie Mellon University. It is a differential drive vehicle with rear caster wheels. A WAAS enabled GPS provides position updates at 1 Hz, which are combined with odometry and the output of an inertial measurement unit (IMU) in an extended Kalman filter (EKF). Sensors include two IR range finders and two stereo heads.

8.3.2 Training and Classification Results

For our dataset, we have extracted and labelled ~16 000 examples drawn from stereo data from eight different locations throughout the United States. Terrain types include forest, meadow, brush and hills. The data were collected

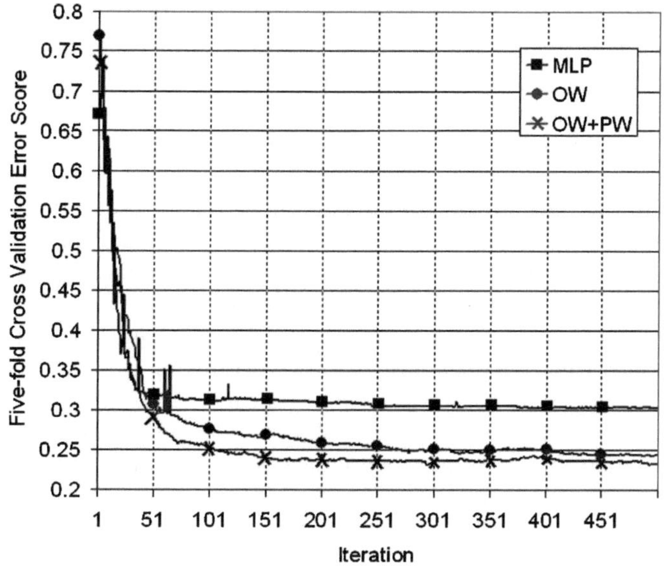

Fig. 8.4. Average error on test set for MLP and convolutional networks

over four seasons and at various times of the day. This dataset includes terrain that is very difficult to discern using height or slope: rocks and logs on the one hand (not traversable), and tall grass (traversable) and bushes (partially traversable) on the other.

To evaluate the effectiveness of convolutional networks operating on 2D histograms, we computed the fivefold cross validation error for the OW and the combined OW+PW classifiers using R^2 range compensation, and compared these results with those from a standard multi-layer perceptron trained on height and slope alone. Results are shown in Fig. 8.4, demonstrating a significant improvement in classification error when using a convolutional neural network on 2D histograms. There is also a noticeable improvement shown by the OW+PW classifier over the single histogram network. To determine the importance of range compensation, we again compared fivefold cross validation error rates, this time for a single network architecture (OW+PW) with varying range compensation applied to its input. There is a mild but noticeable improvement as the degree of range compensation increases, with R^2 range compensation producing the best results.

8.3.3 Field Experiments

We have also tested the effectiveness of the convolutional neural network trained on OW histograms for driving in rugged terrain that is not represented in the network's training data. This network was fielded on the LAGR robot. Only OW histograms were used because of the computational constraints of the platform.

The test course was a winding path approximately 150 m long through woods in western Pennsylvania. The straight line to the goal diverges from the path, bringing the vehicle amongst thin trees and fallen logs. The vehicle navigated the course smoothly, reaching the goal with only one bumper hit, which was on a log no more than 10 cm in height. The average speed on the course was approximately $0.8\,\mathrm{m\,s^{-1}}$. Subsequent runs showed similar performance; the only undetected obstacles were logs less than 10 cm tall.

In January 2007, DARPA conducted official tests of this system at the Southwest Research Institute (SWRI) near San Antonio, Texas. The robot was run on an extremely difficult course, replete with cacti, tall grass, thin trees, overhangs (both passable and impassable), and numerous small logs and fallen branches that represent impassable terrain for this vehicle. Although the vehicle struggled to complete each of its runs, largely due to the presence of low-lying mobility kills, it, nonetheless, avoided many difficult-to-perceive obstacles such as thin trees and overhangs, avoiding all of the former.

Testing at Elgin Air Force Base in Florida in March 2007 produced similar results. The course was 163 m in length through a wooded area that included sandy soil, low grass, pine saplings and trees, brush and azaleas. The system described here was the only one of nine robots to successfully traverse the forest, albeit with one operator intervention when the robot wedged itself on a stump. The traversal took approximately 17 min as the robot searched for a path through the complex terrain.

8.4 Conclusion

We have presented a novel method for terrain classification for mobile robot navigation that eliminates the need for guessing which geometric features are relevant. The system learns to mimic the classification judgments of a human expert rather than applying explicit rules or thresholds to measured features. Its performance compared to robotic systems using the rule-based paradigm indicates that the decade-long reliance on hand-coded methods was misguided. Applying machine learning to robot perception can produce perception systems with greater generality and robustness than attempts to formulate features and rules that apply to all situations.

References

1. Talukder A, Manduchi R, Rankin A, Matthies L (2002) Fast and reliable obstacle detection and segmentation for cross-country navigation. In: Proc. of the IEEE Intelligent Vehicles Symp.
2. LeCun Y, Muller U, Ben J, Cosatto E, Flepp B (2005) Off-road obstacle avoidance through end-to-end learning. In: Advances in Neural Information Processing Systems 18. MIT Press, Cambridge.

3. Vandapel N, Hebert M (2004) Finding organized structures in 3-d ladar data. In: Proc. of the 24th Army Science Conf.
4. LeCun Y, Bottou L, Bengio Y, Haffner P (1998) Gradient-based learning applied to document recognition. Proc. of the IEEE 86:2278–2324.
5. Riedmiller M, Braun H (1993) A direct adaptive method for faster backpropagation learning: the rprop algorithm. In: Proc. of the IEEE Int. Conf. on Neural Networks.

9

A Centre-of-Mass Tracker Integrated Circuit Design in Nanometric CMOS for Robotic Visual Object Position Tracking

S.M. Rezaul Hasan and Johan Potgieter

Centre for Research in Analogue & VLSI microsystem dEsign (CRAVE)
Institute of Technology & Engineering
Massey University, Auckland
New Zealand
hasanmic@massey.ac.nz

Summary. Object position tracking is a well-known problem in robotic vision system design. This chapter proposes a new centre-of-mass(COM) object position tracker integrated circuit design using standard low-cost digital CMOS process technology without the need for a floating gate poly2 layer. The proposed COM tracker is designed using an IBM 130-nm CMOS process with a 1 V supply voltage and can be used for single-chip robotic visual feedback object tracking application. It uses an analogue sampled-data technique and is operated by a two-phase clocking system. The clock frequency can be varied to suit the real-time throughput requirements of the specific robotic object tracking tasks. A 6-input COM detector (easily extendable to 12-input or higher) was simulated and the results for various COM positions was found to be correctly tracked by the circuit. In addition, Monte Carlo simulations were carried out to prove the robustness of the technique.

Keywords: robotic vision, analogue CMOS VLSI, centre-of-mass, robotic object tracking.

9.1 Introduction

Centre-of-mass (COM), the first moment of an object's intensity distribution represents the position of an object. Consequently it has wide applications in robotic vision for visual feedback object tracking [1]. The COM of an object has to be determined in real-time with minimal delay for robotic vision image processing. Compared to digital image processing techniques, analogue processing of image signals provides faster COM response time since no A/D conversion of image signals is required. Yu et al. [2] proposed a COM tracker circuit based on implementation by neuron MOS technology [4]. Although neuron MOS technology simplifies implementation of many robotic and neural

S.M.R. Hasan and J. Potgieter: *A Centre-of-Mass Tracker Integrated Circuit Design in Nanometric CMOS for Robotic Visual Object Position Tracking*, Studies in Computational Intelligence (SCI) **76**, 71–81 (2007)
www.springerlink.com

network circuits [3], it requires extra CMOS processing steps (e.g. POLY2) which is not usually available in many low-cost CMOS process technologies. In this chapter a novel COM detector circuit is proposed which replaces the need for floating gate neuron MOSFETs with the use of a capacitive sampled-data technique following [5]. The proposed circuit uses a 1 V supply voltage and is quite robust to process related threshold voltage variations. The circuit can be extended to any number of analogue image signal inputs and produces a binary (thresholding) COM decision which can be stored as a digital signal in a micro-controller status register. The preliminary work was reported in [7].

9.2 Proposed COM Circuit Architecture

Figure 9.1 shows the basic algorithm to be implemented in each axis for a 2D separable image intensity distribution of an object (e.g. a creature object) for the determination of its COM. Each axis thus requires a separate block of the proposed COM detector. Figure 9.2 shows the circuit architecture of the proposed sampled-data centre-of-mass tracker circuit. It essentially consists of (n+1) capacitive moment-of-intensity circuits and n sampled-data inverter comparators for an n-position COM tracker circuit. The location (position) of the moment-of-intensity for a set of analogue sampled intensities [v_1, v_2, v_3,..., v_n] using capacitive position-scalars [C, 2C, 3C,..., nC] is given by,

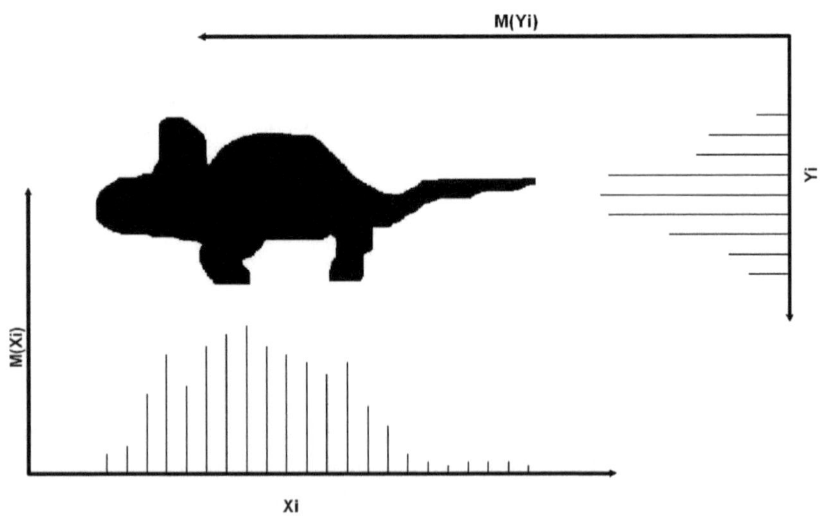

Fig. 9.1. Illustration of the centre-of-mass (COM) detection algorithm. 2D separable image intensity distribution projections in X and Y directions are separately obtained and processed

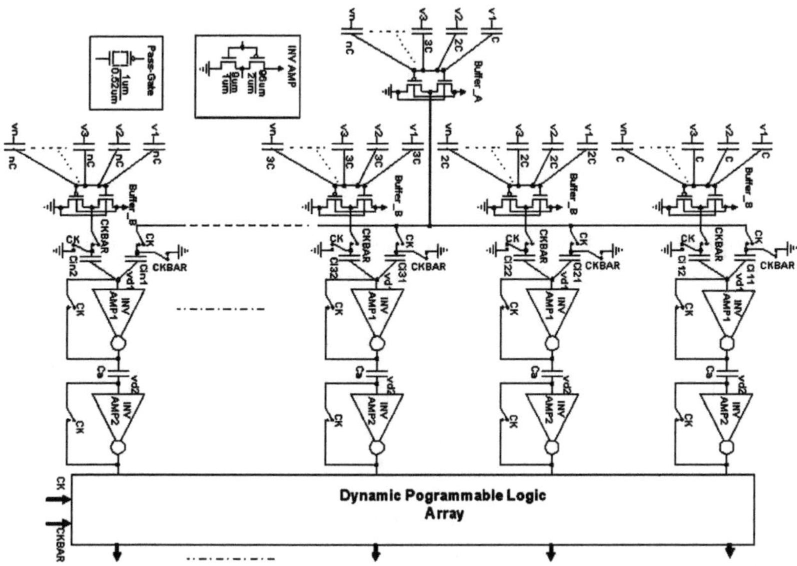

Fig. 9.2. Architecture of the proposed sampled-data real-time centre-of-mass detector circuit for each axis of the 2D separable image

$$C_x = \frac{\sum\limits_{r=1}^{n} rCv_r}{\sum\limits_{r=1}^{n} v_r} \tag{9.1}$$

where C_x is the capacitive COM position-scalar.

Next, (9.1) can be rearranged as,

$$C_x \sum_{r=1}^{n} v_r = \sum_{r=1}^{n} rCv_r \tag{9.2}$$

Now, introducing a proportionality constant, k, (9.2) can be written as,

$$\left[\frac{C_x \sum\limits_{r=1}^{n} v_r}{nC_x}\right] * k * nC_x = \left[\frac{\sum\limits_{r=1}^{n} rCv_r}{\sum\limits_{r=1}^{n} rC}\right] * k * \sum_{r=1}^{n} rC \tag{9.3}$$

Next, dividing both sides by,

$k * nC_x + k * \sum\limits_{r=1}^{n} rC$, and changing sides, we have,

$$\frac{\left[\dfrac{C_x \sum\limits_{r=1}^{n} v_r}{nC_x}\right] * k * nC_x - \left[\dfrac{\sum\limits_{r=1}^{n} rCv_r}{\sum\limits_{r=1}^{n} rC}\right] * k * \sum\limits_{r=1}^{n} rC}{k * nC_x + k * \sum\limits_{r=1}^{n} rC} = 0 \tag{9.4}$$

In the proposed COM circuit, the closest value $C_x = rC$ is sought so that (9.4) holds. So, making this replacement in (9.4), we have,

$$\frac{\left[\dfrac{rC \sum\limits_{r=1}^{n} v_r}{nrC}\right] * k * nrC - \left[\dfrac{\sum\limits_{r=1}^{n} rCv_r}{C * \frac{n(n+1)}{2}}\right] * k * C * \frac{n(n+1)}{2}}{k * nrC + k * C * \frac{n(n+1)}{2}} = M \tag{9.5}$$

where, the specific value r is sought for which the difference M changes sign. The proposed COM tracker is an implementation of the (9.5) with the value r (for which M changes sign) being the COM position detected by the circuit. In the implementation architecture of Fig. 9.2, n sampled data comparison paths are shown (for an n position COM tracker) with the top path (position) for r=1 and the bottom path (position) for r=n. For all the paths,

$$C_{i11} = C_{i21} = C_{i31} = \ldots.. = C_{ir1} = \ldots\ldots = C_{in1} = k * C * \frac{n(n+1)}{2} \tag{9.6}$$

And, for any path r,

$$C_{ir2} = knrC \tag{9.7}$$

Ca is an appropriately chosen value for sampled-data input to the second inverter-amplifier. The rth path (position) source-follower Buffer_B produces the output voltage, $\dfrac{rC \sum\limits_{r=1}^{n} v_r}{nrC}$ = Vr (say). Also, the source-follower Buffer_A produces the output voltage, $\dfrac{\sum\limits_{r=1}^{n} rCv_r}{\sum\limits_{r=1}^{n} rC}$ = Vc(say). The two inputs Vc and Vr along the rth path are sampled during setup phase, PH1 (using clock-signal, CK) and the comparison phase, PH2 (using the clock-bar-signal, CKBAR) intervals, respectively, of a 50% duty cycle global truly single phase clock (TSPC) signal. In a TSPC scheme [6] CKBAR is derived directly by inverting CK instead of using a non-overlapping clock generator, which provides considerable savings in silicon real-estate that would otherwise be consumed by the clock generator circuit. The perturbation signal at the input of the inverter-amplifier 1 (V_{d1}) in the rth path during PH2 interval is due to the difference of two capacitively divided voltages and is then given by,

$$V_{d1} = \frac{(Vc * C_{ir1} - Vr * C_{ir2})}{(C_{ir1} + C_{ir2} + C_{ip})} \tag{9.8}$$

where, C_{ip} is the input parasitic capacitance of the inverter-amplifier. Due to electrostatic coupling of only the weighted difference signal, $(Vc * C_{ir1} - Vr * C_{ir2})$, both the inputs Vc and Vr can individually vary over a wide dynamic range. Both the inverter-amplifiers are self-biased at the common mode voltage (V_{cm}) which is set at half the supply-voltage in order to provide maximum bipolar signal swing at the output. The two inverter-amplifiers are disabled during the setup phase (i.e. during the PH1 interval) via a short-circuit switched by a pass-gate using the CK signal. During the comparison phase (i.e. during the PH2 interval) the perturbation input voltage V_{d1} (given by (9.8)) at the input of the inverter-amplifier 1 is amplified by its gain A1. As a result the perturbation input V_{d2} at the input of the inverter-amplifier 2 in the rth path, which is also due to a capacitive voltage division, is given by,

$$v_{d2} = \left[\frac{(Vc * Cir1 - Vr * Cir2)}{(Cir1 + Cir2 + C_{ip})} * A1\right] * \frac{Ca}{Ca + C_{ip}} \tag{9.9}$$

Finally, this perturbation signal V_{d2} is amplified by the gain A2 of the inverter-amplifier 2 (which is also enabled during the PH2 period). The rth path (position) starting at which the output of the inverter-amplifier 2 switches to opposite polarity compared to the (r−1)th path (position) is the tracked COM position for the input image intensity distribution. A Programmable logic array (PLA) or other simplified logic can be used to flag the COM position using the truth table in Table 9.1. The simple logic implementation using CMOS Nand & Inverter logic gates is shown in Fig. 9.3. The problem

Table 9.1. PLA truth table of the COM tracker output logic

PLA Input	PLA ouput	Comment
011111111111	100000000000	COM @ boundary of object
001111111111	010000000000	COM @ axis_position_2
000111111111	001000000000	COM @ axis_position_3
000011111111	000100000000	COM @ axis_position_4
000001111111	000010000000	COM @ axis_position_5
000000111111	000001000000	COM @ axis_position_6
000000011111	000000100000	COM @ axix_position_7
000000001111	000000010000	COM @ axis_position_8
000000000111	000000001000	COM @ axis_position_9
000000000011	000000000100	COM @ axis_position_10
000000000001	000000000010	COM @ axis_position_11
000000000000	000000000001	COM @ boundary of object

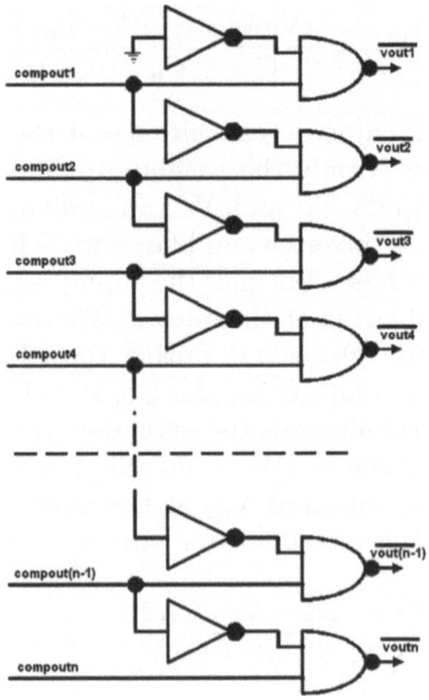

Fig. 9.3. Simple CMOS circuit for the implementation of the PLA truth table of the COM tracker output logic

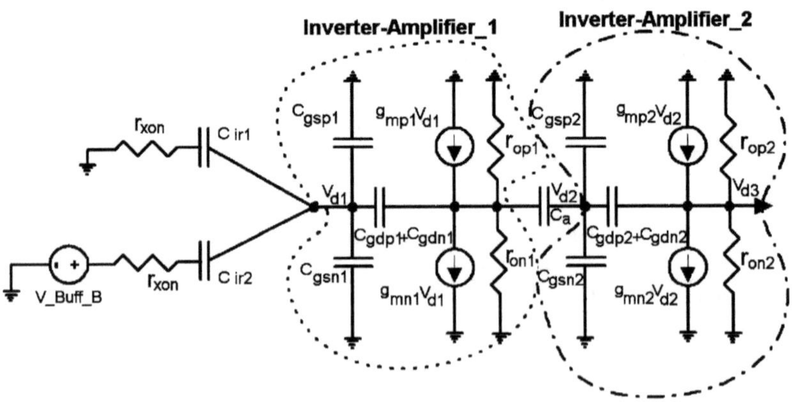

Fig. 9.4. Equivalent circuit of the COM detector's analogue sampled-data signal path during PH2 interval for response time estimation

of charge injection and charge absorption along the pass-transistor controlled signal path can be minimized by using dummy pass gates. The optimum size of the capacitors were determined based on the consideration of response time, parasitic input capacitance of the inverter-amplifiers and charge leakage. The final COM output is available during the comparison phase (PH2) for the input read cycle of any back-end robotic vision logic/instrumentation. Figure 9.4 shows a small signal (perturbation signal) equivalent circuit of the

COM detector's analogue sampled data signal path during the comparison phase (PH2) using standard models for the P-channel MOSFET(PMOS) and N-channel MOSFET(NMOS) devices. Also, standard component notations are used for the resistors, capacitors, voltage and current sources.

9.3 Circuit Simulation and Analysis Results

In order to verify the operation of the proposed novel COM tracker circuit extensive simulations were carried out using the 8M1P 0.13 μm IBM CMOS process technology parameters (BSIM level 49 parameters from a MOSIS wafer lot) using Tanner T-SPICE v.12. The supply voltage was 1.0 V and an input unit capacitance position-scalar of 0.1 pF was used. A 125 kHz clock (time-period = 8 μs) was used as the CK signal for switching the pass gates which is fast enough for many robotic vision systems. Rise-time and fall-time for the clock was set at around 5 ns which can be easily implemented on-chip using standard static CMOS circuit. The inverter-amplifiers are designed using relatively long channel devices for obtaining higher output impedance and resulting higher voltage gain. That way the polarity switch of the 2nd inverter-amplifier's output can be easily detected by the COM tracker output logic (due to a higher noise margin). A six position COM tracker was simulated so that n = 6. Ca was set at 1 pF. k was set to 1, and, the maximum value of capacitance used was 3.6 pF. The circuit design of the pass-gates and the inverter-amplifiers is shown in Fig. 9.2. The device sizes are shown next to the transistors. The tracked COM output is indicated with an active-low signal during PH2 in the COM output logic. Figure 9.5 shows the output of the 6-input COM detector circuit with discrete positions of 0.1 through 0.6, for an exact COM capacitive position-scalar value of 0.46. Figure 9.5a shows the outputs for positions 1, 2, 3, 4 and 6 (all 'HIGH') while Fig. 9.5b shows the output for position 5 ('LOW' during PH2), and, Fig. 9.5c shows the clock PH2 signal. So the nearest COM position 5 (corresponding to the capacitive position-scalar value of 0.5) is tracked by the circuit corresponding to the exact value of 0.46. Similarly, COM position is also correctly tracked for an exact COM of 0.28 as indicated in Fig. 9.6, resulting in a tracked COM position of 3. Next, 100 iterations of Monte Carlo simulations are carried out for process related threshold voltage variation of 25% using a Gaussian distribution function. An exact COM position of 0.38 was applied at the input. The COM was correctly tracked to position 4 as shown in Fig. 9.7. This proves that the proposed architecture is quite robust under process variations in any low-cost digital CMOS technology. The static power dissipated by the COM tracker from a 1 V supply voltage was under 100 μW making it extremely suitable for mobile robotic systems. The response time constraint of the COM tracker circuit can be determined by identifying the major time-constant nodes of the circuit. For this purpose, Fig. 9.4 showing the small signal (perturbation signal) equivalent circuit for the COM detector's signal path (along any rth path)

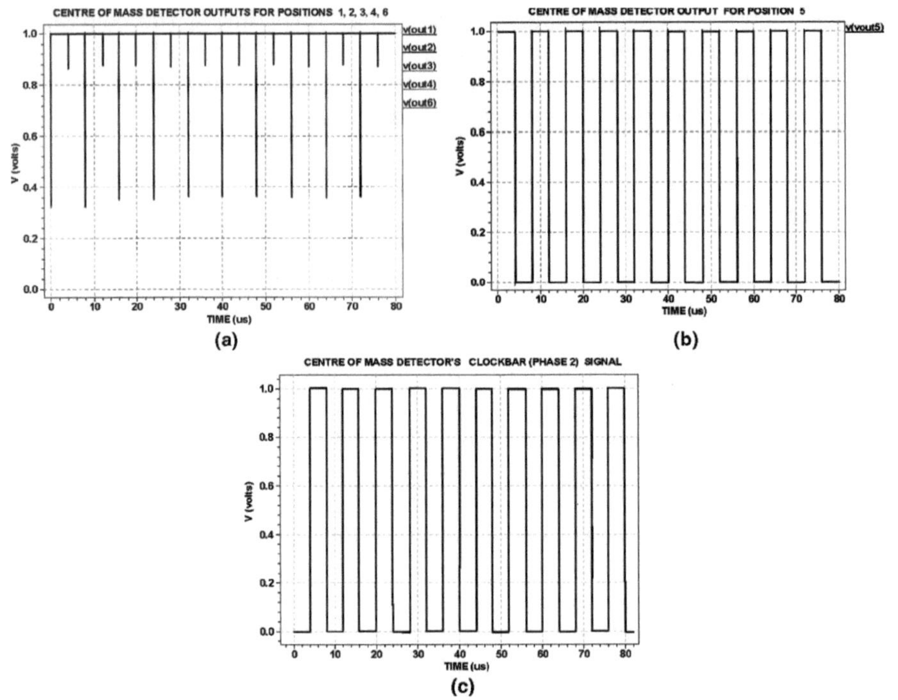

Fig. 9.5. Final output of the 6-input COM detector circuit for a COM of 0.46, (**a**) outputs for positions 1, 2, 3, 4 and 6, (**b**) output for position 5 and (**c**) the clock phase 2(PH2)

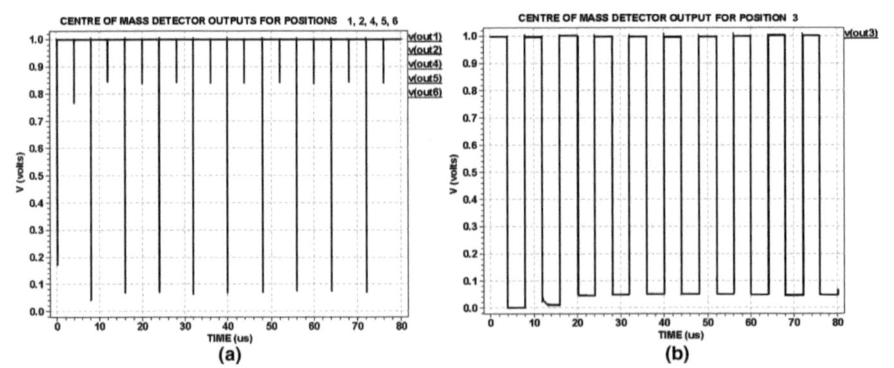

Fig. 9.6. Final output of the 6-input COM detector circuit for a COM of 0.28, (**a**) outputs for positions 1, 2, 4, 5 and 6, (**b**) output for position 3

during the comparison phase can be utilized. The dotted lines enclose equivalent circuits for the sub-block components. Here r_{xon} is the finite on-resistance of the transmission-gate switches (ideally should be very small). All other symbols and notations have their usual meaning for standard small signal

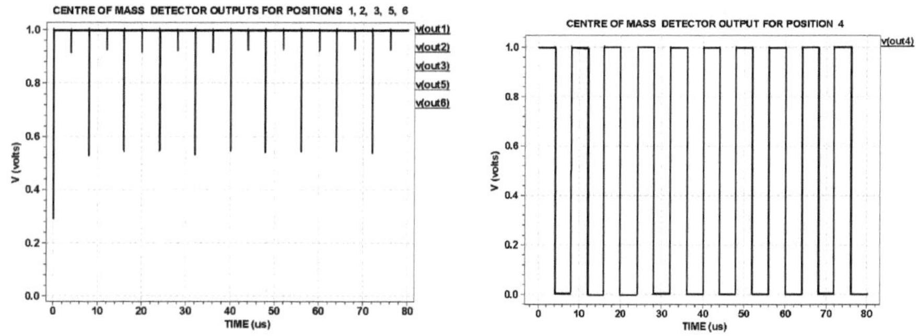

(a) (b)

Fig. 9.7. Final output of the 6-input COM detector circuit for a COM of 0.38, (**a**) outputs for positions 1, 2, 3, 5 and 6, (**b**) output for position 4, considering threshold voltage variation by over 25% after 100 iterations of Monte Carlo simulations

MOSFET model as follows: C_{gsp} is the PMOS gate-to-source capacitance, C_{gsn} is the NMOS gate-to-source capacitance, C_{gdp} is the PMOS gate-to-drain capacitance, C_{gdn} is the NMOS gate-to-drain capacitance, r_{op} is the PMOS output impedance, r_{on} is the NMOS output impedance, g_{mp} is the PMOS transconductance and finally g_{mn} is the NMOS transconductance. The dominant time constants are at the *outputs* of the inverter-amplifier 1 and the inverter-amplifier 2. The time constant (τ_1) at the *output* of inverter-amplifier 1 is given by (including the miller equivalent capacitance due to the gate-to-drain capacitances of inverter-amplifier 2),

$$\tau_1 \approx \left[\frac{r_{op1} r_{on1}}{r_{op1} + r_{on1}}\right] * \left[\begin{array}{c}(C_{gdp1} + C_{gdn1}) + \\ \frac{Ca*(C_{gsp2}+C_{gsn2}+g_{mp2}r_{op2}C_{gdp2}+g_{mn2}r_{on2}C_{gdn2})}{Ca+(C_{gsp2}+C_{gsn2}+g_{mp2}r_{op2}C_{gdp2}+g_{mn2}r_{on2}C_{gdn2})}\end{array}\right] \quad (9.10)$$

where, the overall resistance at this node is the parallel combination of the NMOS and PMOS output impedances of the inverter-amplifier 1, r_{op1} and r_{on1}. Also, the sampling capacitor Ca is in series with the parallel combination of the capacitances C_{gsn2}, C_{gsp2} and the miller multiplied capacitances $g_{mp2}r_{op2}C_{gdp2}$ and $g_{mn2}r_{on2}C_{gdn2}$. Also, this equivalent capacitance is in parallel with the sum of the capacitances C_{gdp1} and C_{gdn1}.

Next, the time constant (τ_2) at the *output* of the inverter-amplifier 2 is given by,

$$\tau_2 \approx \frac{r_{op2} r_{on2}}{(r_{op2} + r_{on2})} * (C_{gsnL} + C_{gspL} + C_{gdn2} + C_{gdp2}) \quad (9.11)$$

where, the overall resistance at this node is the parallel combination of the NMOS and PMOS output impedances of the inverter-amplifier 2, r_{op2} and r_{on2}. Also, C_{gsnL} and C_{gspL} are the logic gate input capacitances which are in parallel with the miller equivalent capacitances C_{gdp2} and C_{gdn2} at the output of inverter-amplifier 2. The overall −3dB clock bandwidth for the COM tracker is then given by,

$$\omega_{3dBClock} = \frac{1}{\tau_1 + \tau_2} \tag{9.12}$$

Inspecting the (9.10) and (9.11), it is clear that the response is dominated by the time constant τ_1 at the *output* of the inverter-amplifier 1. Since Ca (1 pF in this design) is large compared to the parasitic device capacitances and the miller multiplied capacitances by an order of magnitude, the −3dB clock bandwidth of the comparator can then be approximated by,

$$\omega_{3dBClock} \approx \frac{1}{\frac{r_{op1} r_{on1}}{(r_{op1}+r_{on1})} * C_{tot}} \tag{9.13}$$

where,

$$\begin{aligned} C_{tot} &= C_{gdp1} + C_{gdn1} + \\ &C_{gsp2} + C_{gsn2} + g_{mn2} r_{on2} C_{gdn2} + g_{mp2} r_{op2} C_{gdp2} \end{aligned} \tag{9.14}$$

The −3dB clock bandwidth thus depends on the size $(W*L)$ of the transistors channel-area and the bias drain current (I_D). As the intrinsic inverter-amplifier gain $(g_m r_o)$ is $\propto \sqrt{\frac{W*L}{I_D}}$ and output impedance r_o is $\propto \frac{1}{I_D}$, for a given drain current, I_D, the response time will trade with the output signal swing (using higher voltage gains) driving the logic gates, and hence the noise margins.

9.4 Conclusion

A low-voltage centre-of-mass tracker circuit design using only 1 V supply voltage has been achieved. The significance of this circuit compared to other recent COM tracker circuits is that it does not require floating-gate POLY2 process layer and can be conveniently fabricated in any standard low-cost digital CMOS process technology. Also, although a 6-position COM tracker is presented in this chapter, the novel architecture is very modular and can be easily extended to any number of COM positions. Thus, the precision achievable by the COM tracker circuit at such low supply voltage makes it favourable for applications in mobile robot visual object tracking system.

References

1. I. Ishii, Y. Nakabo, and M. Ishikawa, " Target tracking algorithms for 1ms visual feedback system using massively parallel processing", *Proceedings IEEE Int. Conf. Robotics and Automation*, vol. 3, pp. 2309–2314 (1996).
2. N. M. Yu, T. Shibata, and T. Ohmi, "A real-time center-of-mass tracker circuit implemented by neuron MOS technology", *IEEE Trans. Circuits & Systems-II: Analog and Digital Signal Processing*, vol. 45, no. 4. pp. 495–503 (1998).

3. K. Nakada, T. Asai, and Y. Amemiya, "Analog CMOS implementation of a CNN-based locomotion controller with floating gate devices", *IEEE Trans. Circuits & Systems-I: Regular Papers*, vol. 52, no. 6, pp. 1095–1103 (2005).
4. W. Weber, S. J. Prange, R. Thewes, E. Wohlrab, and A. Luck, "On the application of the neuron MOS transistor principle for modern VLSI design", *IEEE Trans. on Electron Devices*, vol. 43, no. 10, pp. 1700–1708 (1996).
5. X. Li, Y. Yang, and Z. Zhu, "A CMOS 8-bit two-step A/D converter with low power consumption ", *Proceedings IEEE Int. Workshop VLSI Design & Video Tech., Suzhou, China, May 28–30, 2005*, pp. 44–47 (2005).
6. Y. J. Ren, I. Karlsson, C. Svensson, "A true single- phase-clock dynamic CMOS circuit technique", *IEEE J. Solid-State Circuits*, Vol. 22, no. 5, pp. 899–901 (1987).
7. S. M. Rezaul Hasan and J. Potgieter, " A novel CMOS sample-data centre-of-mass tracker circuit for robotic visual feedback object tracking", *Proceedings the 3rd International conference on Autonomous Robots and agents*, Palmerston North, New Zealand, pp. 213–218, (2006).

10

Non-Iterative Vision-Based Interpolation of 3D Laser Scans

Henrik Andreasson[1], Rudolph Triebel[2], and Achim Lilienthal[1]

[1] AASS, Dept. of Technology, Örebro University, Sweden
henrik.andreasson@tech.oru.se, achim.lilienthal@tech.oru.se
[2] Autonomous Systems Lab, ETH Zürich, Switzerland
rudolph.triebel@mavt.ethz.ch

Summary. 3D range sensors, particularly 3D laser range scanners, enjoy a rising popularity and are used nowadays for many different applications. The resolution 3D range sensors provide in the image plane is typically much lower than the resolution of a modern colour camera. In this chapter we focus on methods to derive a high-resolution depth image from a low-resolution 3D range sensor and a colour image. The main idea is to use colour similarity as an indication of depth similarity, based on the observation that depth discontinuities in the scene often correspond to colour or brightness changes in the camera image. We present five interpolation methods and compare them with an independently proposed method based on Markov random fields. The proposed algorithms are non-iterative and include a parameter-free vision-based interpolation method. In contrast to previous work, we present ground truth evaluation with real world data and analyse both indoor and outdoor data.

Keywords: 3D range sensor, laser range scanner, vision-based depth interpolation, 3D vision.

10.1 Introduction

3D range sensors are getting more and more common and are found in many different areas. A large research area deals with acquiring accurate and very dense 3D models, potential application domains include documenting cultural heritage [1], excavation sites and mapping of underground mines [2]. A lot of work has been done in which textural information obtained from a camera is added to the 3D data. For example, Sequeira et al. [3] present a system that creates textured 3D models of indoor environments using a 3D laser range sensor and a camera. Früh and Zakhor [4] generate photo-realistic 3D reconstructions from urban scenes by combining aerial images with textured 3D data acquired with a laser range scanner and a camera mounted on a vehicle.

H. Andreasson et al.: *Non-Iterative Vision-Based Interpolation of 3D Laser Scans*, Studies in Computational Intelligence (SCI) **76**, 83–90 (2007)
www.springerlink.com

Fig. 10.1. *Left*: Image intensities plotted with the resolution of the 3D scanner. The laser range readings were projected onto the right image and the closest pixel regions were set to the intensity of the projected pixel for better visualisation. *Middle*: Calibration board used for finding the external parameters of the camera, with a chess board texture and reflective tape (*grey border*) to locate the board in 3D using the remission/intensity values from the laser scanner. *Right*: Natural neighbours $R_1 \dots R_5$ of R_i^*. The interpolated weight of each natural neighbour R_i is proportional to the size of the area which contains the points Voronoi cell and the cell generated by R_j^*. For example the nearest neighbour R_1 will have influence based upon the area of A_1

In most of the approaches that use a range scanner and a camera, the vision sensor is not actively used during the creation of the model. Instead vision data are only used in the last step to add texture to the extracted model. An exception is the work by Haala and Alshawabkeh [5], in which the camera is used to add line features detected in the images into the created model.

To add a feature obtained with a camera to the point cloud obtained with a laser range scanner, it is required to find the mapping of the 3D laser points onto pixel co-ordinates in the image. If the focus instead lies on using the camera as an active source of information which is considered in this chapter, the fusing part in addition addresses the question of how to estimate a 3D position for each (sub) pixel in the image. The resolution that the range sensor can provide is much lower than those obtained with a modern colour camera. This can be seen by comparing left of Fig. 10.1, created by assigning the intensity value of the projected laser point to its closest neighbours, with the corresponding colour image in middle of Fig. 10.1.

This chapter is a shortened version of [6].

10.2 Suggested Vision-Based Interpolation Approaches

The main idea is to interpolate low-resolution range data provided by a 3D laser range scanner under the assumption that depth discontinuities in the scene often correspond to colour or brightness changes in the camera image of the scene.

For the problem under consideration, a set of N laser range measurements $r_1 \ldots r_N$ is given where each measurement $r_i = (\theta_i, \pi_i, r_i)$ contains a tilt angle θ_i, a pan angle π_i and a range reading r_i corresponding to 3D Euclidean coordinates (x_i, y_i, z_i).

The image data consists of a set of image pixels $\mathrm{P}_j = (X_j, Y_j, C_j)$, where X_j, Y_j are the pixel co-ordinates and $C_j = (C_j^1, C_j^2, C_j^3)$ is a three-channel colour value. By projecting a laser range measurement r_i onto the image plane, a projected laser range reading $\mathrm{R}_i = (X_i, Y_i, r_i, (C_i^1, C_i^2, C_i^3))$ is obtained, which associates a range reading r_i with the coordinates and the colour of an image pixel. An image showing the projected intensities can be seen in Fig. 10.1, where the closest pixel regions are set to the intensity of the projected pixel for better visualisation. The interpolation problem can now be stated for a given pixel P_j and a set of projected laser range readings R, as to estimate the interpolated range reading r_j^* as accurately as possible. Hence we denote an interpolated point $\mathrm{R}_j^* = (X_j, Y_j, r_j^*, C_j^1, C_j^2, C_j^3)$.

Five different interpolation techniques are described in this section and compared with the MRF approach described in Sect. 10.3.

10.2.1 Nearest Range Reading (NR)

Given a pixel P_j, the interpolated range reading r_j^* is assigned to the laser range reading r_i corresponding to the projected laser range reading R_i which has the highest likelihood p given as

$$p(\mathrm{P}_j, \mathrm{R}_i) \propto e^{-\frac{(X_j - X_i)^2 + (Y_j - Y_i)^2}{\sigma^2}}, \tag{10.1}$$

where σ is the point distribution variance. Hence, the range reading of the closest point (regarding pixel distance) will be selected.

10.2.2 Nearest Range Reading Considering Colour (NRC)

This method is an extension of the NR method using colour information in addition. Given a pixel P_j, the interpolated range reading r_j^* is assigned to the range value r_i of the projected laser range reading R_i which has the highest likelihood p given as

$$p(\mathrm{P}_j, \mathrm{R}_i) \propto e^{-\frac{(X_j - X_i)^2 + (Y_j - Y_i)^2}{\sigma_p^2} - \frac{||C_j - C_i||^2}{\sigma_c^2}}, \tag{10.2}$$

where σ_p and σ_c is the variance for the pixel point and the colour, respectively.

10.2.3 Multi-Linear Interpolation (MLI)

Given a set of projected laser range readings $\mathrm{R}_1 \ldots \mathrm{R}_N$, a Voronoi diagram V is created by using their corresponding pixel co-ordinates $[X, Y]_{1 \ldots N}$. The natural neighbours NN to an interpolated point R_j^* are the points in V, which

Fig. 10.2. From *left* to *right*: Depth image generated with the NR method. Depth image generated with the NRC method, small details are now visible, note that a depth image generated from a similar viewpoint as the laser range scanner makes it very difficult to see flaws of the interpolation algorithm. MLI method. LIC method

Voronoi cell would be affected if R_j^* is added to the Voronoi diagram, see Fig. 10.1. By inserting R_j^* we can obtain the areas $A_{1...n}$ of the intersection between the Voronoi cell due to R_j^* and the Voronoi cell of R_i before inserting R_j^* and the area $A_{R_j^*}$ as a normalisation factor. The weight of the natural neighbour R_i is calculated as

$$w_i(R_j^*) = \frac{A_i}{A_{R_j^*}}. \tag{10.3}$$

The interpolated range reading r_j^* is then calculated as

$$r_j^* = \sum_{i \in NN(R_j^*)} w_i r_i. \tag{10.4}$$

This interpolation approach is linear [7]. One disadvantage is that nearest neighbourhood can only be calculated within the convex hull of the scan-points projected to the image. However, this is not considered as a problem since the convex hull encloses almost the whole image, see Fig. 10.2.

10.2.4 Multi-Linear Interpolation Considering Colour (LIC)

To fuse colour information with the MLI approach introduced in the previous subsection, the areas A_{R_i} and $A_{R_j^*}$ are combined with colour weights $w_{1...n}^c$ for each natural neighbour based on spatial distance in colour space.

Similar as in Sect. 10.2.2, a colour variance σ_c is used:

$$w_i^c(R_j^*) = e^{-\frac{||C_i - C_j||^2}{\sigma_c^2}}. \tag{10.5}$$

The colour-based interpolated range reading estimation is then done with

$$r_j^* = \sum_{i \in NN(R_j)} \frac{w_i w_i^c}{W^c} r_i \tag{10.6}$$

where $W^c = \sum_{i=1}^{n} w_i^c$ is used as a normalisation factor.

10.2.5 Parameter-Free Multi-Linear Interpolation Considering Colour (PLIC)

One major drawback of the methods presented so far and the approach presented in the related work section is that they depend on parameters such as σ_c, for example. To avoid the need to specify colour variances, the intersection area A_{R_i} defined in Sect. 10.2.3 is used to compute a colour variance estimate for each nearest neighbour point R_i as

$$\sigma_{c_i} = \frac{1}{n_i - 1} \sum_{j \in A_i} ||\mu_i - C_j||^2, \tag{10.7}$$

where $\mu_i = \frac{1}{n_i} \sum_{j \in A_i} C_j$ and n_i is the number of pixel points within the region A_i. σ_{c_i} is then used in (10.5).

This results in an adaptive adjustment of the weight of each point. In case of a large variance of the local surface texture, colour similarity will have less impact on the weight w_i.

10.3 Related Work

To our knowledge, the only work using vision for interpolation of 3D laser data is [8] where a Markov random field (MRF) framework is used.

The method works by iteratively minimising two constraints: ψ stating that the raw laser data and the surrounding estimated depths should be similar and ϕ stating that the depth estimates close to each other with a similar colour should also have similar depths.

$$\psi = \sum_{i \in N} k(r_i^* - r_i)^2, \tag{10.8}$$

where k is a constant and the sum runs over the set of N positions which contain a laser range reading r_i and r_i^* is the interpolated range reading for position i. The second constraint is given as

$$\phi = \sum_{i} \sum_{j \in NN(i)} e^{(-c||C_i - C_j||^2)} (r_i^* - r_j^*)^2, \tag{10.9}$$

where c is a constant, C is the pixel colour and $NN(i)$ are the neighbourhood pixels around position i. The function to be minimised is the sum $\psi + \phi$.

10.4 Evaluation

All datasets D were divided into two equally sized parts D_1 and D_2. One dataset, D_1, is used for interpolation and D_2 is used as the ground truth where each laser range measurement is projected to image co-ordinates. Hence for

Fig. 10.3. *Left*: The third indoor evaluation scan, $Indoor_3$. *Middle*: Scans taken in winter time with some snow containing $Outdoor_1 - Outdoor_3$. *Right*: Our outdoor robot with the SICK LMS scanner and a colour CCD camera mounted on a pan tile unit from Amtec, that were used in the experiments. The close-up part shows the displacement between the camera and the laser which causes parallax errors

each ground truth point R_i we have the pixel positions $[X, Y]_i$ and the range r_i. The pixel position $[X, Y]_i$ is used as input to the interpolation algorithm and the range r_i is used as the ground truth. The performance of the interpolation algorithms is analysed based on the difference between the interpolated range r_i^* and the range r_i from the ground truth.

10.5 Experimental Setup

The scanner used is a 2D SICK LMS-200 mounted together with a 1 MegaPixel (1280×960) colour CCD camera on a pan-tilt unit from Amtec where the displacement between the optical axis is approx. 0.2 m. The scanner is located on our outdoor robot, see Fig. 10.3, a P3-AT from ActivMedia. The angular resolution of the laser scanner is 0.5°. Half of the readings were used as ground truth, so the resolution for the points used for interpolation is 1°.

10.6 Results

In all experiments the colour variance $\sigma_c = 0.05$ and the pixel distance variance $\sigma_d = 10\,\text{mm}$ were used, which were found empirically. The parameters used within the MRF approach described in Sect. 10.3, where obtained by extensive empirical testing and were set to $k = 2$ and $C = 10$. The optimisation method used for this method was the conjugate gradient method described in [9] and the initial depths were estimated with the NR method. In all experiments the full resolution (1280×960) of the camera image was used.

All the interpolation algorithms described in this chapter were tested on real data consisting of three indoor and outdoor scans. The outdoor scans were taken in winter time with snow, which presents the additional challenge

Table 10.1. Results from $Indoor_1$, $Indoor_2$ and $Indoor_3$ datasets

	NR	NRC	MLI	LIC	PLIC	MRF
$Indoor_1$	0.065	0.054	0.052	0.048	0.049	0.048
$Indoor_2$	0.123	0.134	0.109	0.107	0.109	0.106
$Indoor_3$	0.088	0.072	0.067	0.060	0.060	0.067

Table 10.2. Results from $Outdoor_1$, $Outdoor_2$ and $Outdoor_3$ datasets

	NR	NRC	MLI	LIC	PLIC	MRF
$Outdoor_1$	0.067	0.068	0.056	0.059	0.054	0.054
$Outdoor_2$	0.219	0.294	0.235	0.322	0.275	0.218
$Outdoor_3$	0.526	0.584	0.522	0.574	0.500	0.498

that most of the points in the scene have very similar colours. The results are summarised in Tables 10.1 and 10.2, which show the mean error with respect to the ground truth.

For the indoor datasets, which comprise many planar structures, the lowest mean error was found with the multi-linear interpolation methods, particularly LIC and PLIC, and MRF interpolation. LIC and PLIC produced less (but larger) outliers.

With the outdoor data the results obtained were more diverse. For the dataset $Outdoor_1$, which contains some planar structures, a similar result as in the case of the indoor data was observed. For datasets with a very small portion of planar structures such as $Outdoor_2$ and $Outdoor_3$, the mean error was generally much higher and the MRF method performed slightly better compared to the multi-linear interpolation methods. This is likely due to the absence of planar surfaces and the strong similarity of the colours in the image recorded at winter time. It is noteworthy that in this case, the nearest neighbour interpolation method *without* considering colour (NR) performed as good as MRF. The interpolation accuracy of the parameter-free PLIC method was always better or comparable to the parameterised method LIC.

10.7 Conclusions

This chapter is concerned with methods to derive a high-resolution depth image from a low-resolution 3D range sensor and a colour image. We suggest five interpolation methods and compare them with an alternative method proposed by Diebel and Thrun [8]. In contrast to previous work, we present ground truth evaluation with simulated and real world data and analyse both indoor and outdoor data. The results of this evaluation do not allow to single out one particular interpolation method that provides a distinctly superior interpolation accuracy, indicating that the best interpolation method depends on the content of the scene. Altogether, the MRF method proposed in [8] and the PLIC method proposed in this chapter provided the best interpolation performance. While providing basically the same level of interpolation accuracy as the MRF approach, the PLIC method has the advantage that it is a parameter-free and non-iterative method, i.e. that a certain processing time can be guaranteed. One advantage of the proposed methods is that depth estimates can be obtained without calculating a full depth image. For example if interpolation points are extracted in the image using a vision-based method (i.e. feature extraction), we can directly obtain a depth estimate for each feature.

References

1. M. Levoy, K. Pulli, B. Curless, S. Rusinkiewicz, D. Koller, L. Pereira, M. Ginzton, S. Anderson, J. Davis, J. Ginsberg, J. Shade, and D. Fulk. The digital michelangelo project: 3D scanning of large statues. In Kurt Akeley, editor, *Siggraph 2000, Computer Graphics Proceedings*, pages 131–144. ACM Press/ACM SIGGRAPH/ Addison Wesley Longman, 2000.
2. S. Thrun, D. Hähnel, D. Ferguson, M. Montemerlo, R. Triebel, W. Burgard, C. Baker, Z. Omohundro, S. Thayer, and W. Whittaker. A system for volumetric robotic mapping of abandoned mines. In *ICRA*, pages 4270–4275, 2003.
3. V. Sequeira, J. Goncalves, and M.I. Ribeiro. 3d reconstruction of indoor environments. In *Proc. of ICIP*, pages 405–408, Lausanne, Switzerland, 1996.
4. C. Früh and A. Zakhor. 3D model generation for cities using aerial photographs and ground level laser scans. In *CVPR*, pages 31–38, Hawaii, USA, 2001.
5. N. Haala and Y. Alshawabkeh. Application of photogrammetric techniques for heritage documentation. In *2nd Int. Conf. on Science & Technology in Archaeology & Conservation*, Amman, Jordan, 2003.
6. H. Andreasson, R. Triebel, and A. Lilienthal. Vision-based interpolation of 3d laser scans. In *Proc. of ICARA*, pages 469–474, 2006.
7. R. Sibson. *A brief description of natural neighbour interpolation*, pages 21–36. John Wiley & Sons, Chichester, 1981.
8. J. Diebel and S. Thrun. An application of markov random fields to range sensing. In Y. Weiss, B. Schölkopf, and J. Platt, editors, *Advances in Neural Information Processing Systems 18*, pages 291–298. MIT Press, Cambridge, MA, 2006.
9. William H. Press, Brian P. Flannery, Saul A. Teukolsky, and William T. Vetterling. *Numerical Recipes: The Art of Scientific Computing.* Cambridge University Press, Cambridge (UK) and New York, 2nd edition, 1992.

11

RoboSim: A Multimode 3D Simulator for Studying Mobile Robot Co-Operation

G. Seet, S.K. Sim, W.C. Pang, Z.X. Wu, and T. Asokan

Robotics Research Centre
School of Mechanical & Aerospace Engineering
Nanyang Technological University, Singapore
MGLSEET@ntu.edu.sg

Summary. The design and development of an interactive 3D robotic simulator, as a platform for the study of co-operative robot behaviour, is described in this chapter. The simulator provides a visually realistic 3D world with simulated robot agents and geographical terrain, for the purpose of evaluating the performance of robots in various scenarios. The simulator was developed in C++ using open source packages, resulting in a modular and cost effective package. The modelling and simulation of robotic vehicles, terrains, sensors and behaviours are presented. Artificial Intelligent agent behaviours in co-operative robotic scenarios were developed and tested using the simulator, and the results presented.

Keywords: Robotics, simulator, co-operative behaviour, artificial agents, multimode.

11.1 Introduction

Simulations [1–4] are frequently used to test the performance of robots in different operating scenarios before deploying them to real tasks. One of the main reasons for this is the cost involved in the actual deployment, as well as the difficulties in executing hundreds of trials using real physical robots. Over years, significant advancements in computer graphics, processor capability, networking and artificial intelligence (AI) have created exciting new options in simulation technology. In this chapter, the focus is on simulators that attempt to model the real world in sufficient level of detail, network distributed and provides a platform for developing and testing algorithms for multiple mobile robots.

The research on the networked simulators has been on going and one of the earliest network simulators is the SIMNET [7] project funded by DARPA to investigate the feasibility of creating a real-time distributed simulator for combat simulation. Distributed interactive simulation (DIS [8]), a follow-on to

G. Seet et al.: *RoboSim: A Multimode 3D Simulator for Studying Mobile Robot Co-Operation*, Studies in Computational Intelligence (SCI) **76**, 91–99 (2007)
www.springerlink.com

the SIMNET system, allowed any type of player or machine complying with its protocol to participate in the simulation. In the academic community, Swedish Institute of Computer Science had developed a Distributed interactive virtual environment (DIVE [9]), an internet-based multi-user virtual reality system. The above-mentioned research works, however, had placed little focus on how realistic these simulations are when compared to the real world. The dynamics in such simulations are only approximate at best. Hence, these simulators lack sufficient accuracy to transfer developed algorithms from simulation to reality with confidence.

Other simulators, such as Gazebo [3] and Webots [4], may provide a dynamically accurate motion, a variety of actuator primitives and support a wide range of sensor suites, but at most time, these simulators are designed to simulate a small population, and compromise the graphical realism, especially for large outdoor environments with trees, buildings and roads.

RoboSim is a piece of work, developed at the Nanyang Technological University, Robotics Research Centre, attempting to provide a generic robotic simulator application, supporting networked multi-users, multi-modes, with realistic terrene environment, sensors and robots modelling. It also provides the ability to test scenarios and port algorithms between the real and virtual environments. Different operating modes like tele-operation, semi-autonomous and autonomous operations are built into RoboSim, giving it a multi-mode simulation capability.

The three-dimensional terrain, environment and entities provide the operator a high level of realism in scenario simulations and in the development of deployment strategy. Inclusion of geo-specific imagery and Digital elevation maps (DEM), for terrain modelling, helps to model and simulate any specific operating terrain as required. The sky, terrain and weather conditions are also changed according to time or humans' desire to demonstrate the effects environment have on different sensors and robots during the simulation.

In RoboSim, different types of robots, e.g. ATRV-JR and Pioneer, are simulated. Different sensors and payloads are modelled as plug-ins for a modular design, and facilitate integration of additional components. RoboSim has been created on the Linux operating system together with some open source graphics libraries [5, 6]. The simulator unifies the software development, control operation and behaviour of real and virtual robots. RoboSim is also designed to be network distributed. Simulation and rendering can be done on different machines connected through a network.

11.2 System Structure

11.2.1 System Components

The three major components of the simulator are Networked Collaborative Virtual Environment (NCVE), entity simulator and sensor module. NCVE functions as a virtual environment simulator with network features. Simulation and rendering can be done on a single, or on different, machines connected through a network.

The entity simulator is responsible for simulating the behaviour/performance of entities like robot vehicles. In the NCVE, each host running RoboSim has two type of entity: local and remote. Local entities run on the local system and broadcast their status to other system. Remote entity serves as a duplicated copy of those entities running on other remote system. Computations are only done for local entity while remote entities are updated only upon receiving update information.

The sensor module simulates the physical characteristics of each type of sensor, including laser scanner, differential GPS, electronic compass, inclinometer and specialized payload, together with their behaviour at various levels of detail and conditions. Environment conditions usually have an effect on some sensor characteristics.

11.2.2 System Configuration

The overall system configuration is identified as shown in Fig. 11.1. There are three distinct layers at the simulator level: Rendering layer, Application layer and AI layer. The rendering layer is responsible for rendering graphical models of simulated environment using hardware accelerated OpenGL script. The AI layer consists of different AI behaviours and is designed to be a separate module and run independent of the simulator. This option was chosen due to its complexity and high computational requirement. Inputs to the AI layer, like data for some sensors, are however collected from the entity running in the simulator. Control commands, from the AI layer, are sent back to the entity. Most of the components are developed as plug-ins to the simulator. This structure provides a clearer interface and ease of software maintenance.

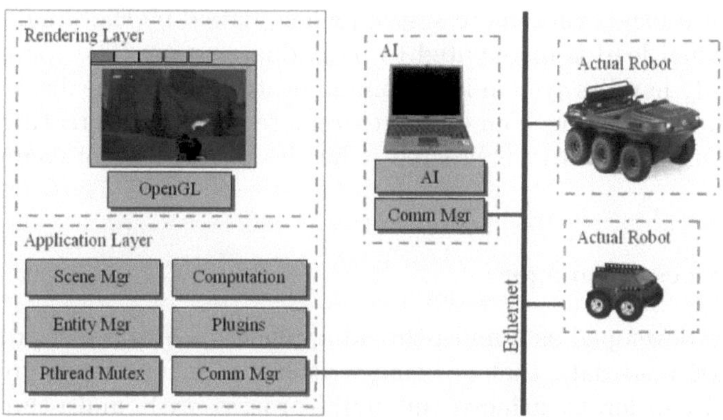

Fig. 11.1. System configuration

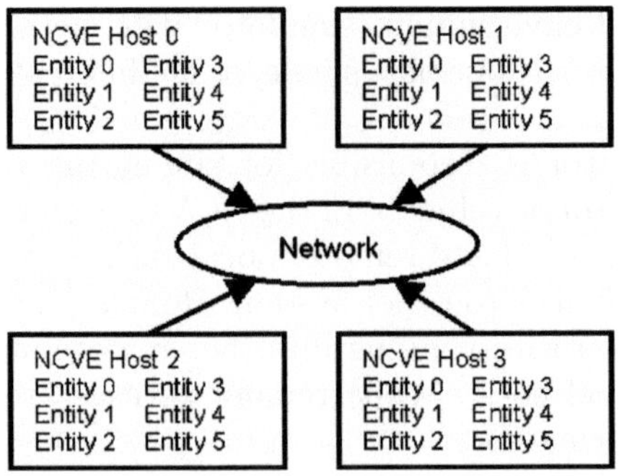

Fig. 11.2. Robotic simulator network layout

The application layer consists of the scene manager, entity manager, the process manager and network communication manager.

The scene manager organizes the entire object in the simulation to provide realistic scene and environments at the lowest possible processing cost. It incorporates advanced 3D rendering techniques like view culling and the level of detail (LOD). Each entity in the simulator has a graphical model, a collision model and a kinematics model. The entity is simulated as an independent computational unit.

11.2.3 Communication Manager and Network Layout

The communication manager implements a TCP/IP network interface. We have used two communication protocols in our network: UDP/IP and TCP/IP. Critical information like tasks assignment is sent through TCP whilst information like robot status, payload status is through the UDP protocol.

Figure 11.2 shows the network layout of the robotic simulator running on various hosts connected by a TCP/IP network. All the hosts differ only as being either a local or a remote system, when viewed by the simulated entity. Computational load is distributed to individual system. The communication manager only handles system level communication tasks, like discovering running hosts, establishing connection to remote hosts and creating entity to simulate. Once the entity is created it has its own means of communication capability.

11.2.4 Process Manager

RoboSim is developed as a multi-thread application. A parallel process, however, cannot read data that is being written by an upstream process, concurrently. In order to achieve full multi-process with multi-thread safety, a large overhead of data buffering, protection and synchronization need to be added to the system processes. General multi-threaded read and write

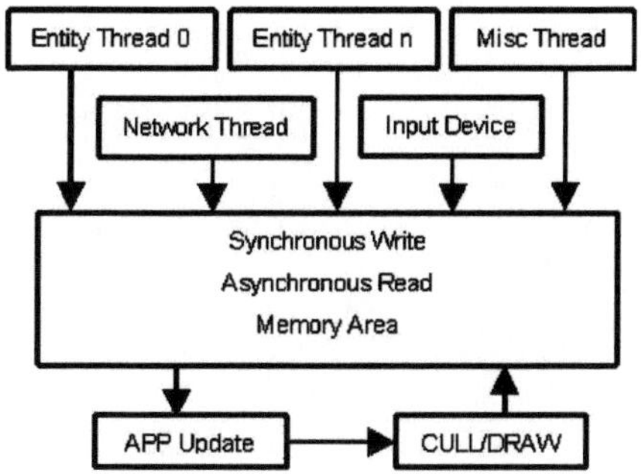

Fig. 11.3. Robotic simulator process model

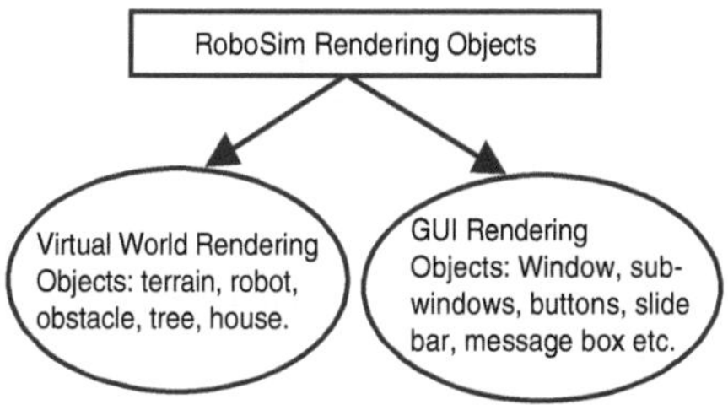

Fig. 11.4. Robotic simulator rendering object

access is computational inefficient and complicates its design. In RoboSim, we employed a Synchronous Write, Asynchronous Read (SWAR) approach (Fig. 11.3). Shared data can only be written by a single thread and read by multiple threads. On the top portion of Fig. 11.3 is shown various threads like entity threads and network threads. These are run independently.

11.3 Graphical User Interface

The basic GUI design concept, of the simulator, is to develop a separate and independent class for the GUI package. As shown in Fig. 11.4, the GUI works as a separate rendering object. This allows the concurrent development of the virtual world and GUI, and testing to be performed separately.

The GUI is based on OpenGL, a platform similar to that for graphical rendering. This maximizes the portability of the software and provides a unified programming interface. A variety of input devices such as mouse, keyboard and joysticks are provided to allow different functionalities for the operator to interact efficiently with the simulator and control the activities. Figure 11.5 shows an overview of the simulator running multiple robots, together with its GUI. The GUI includes the main menu, map view window, camera view and robot information windows.

Fig. 11.5. Robotic simulator overview

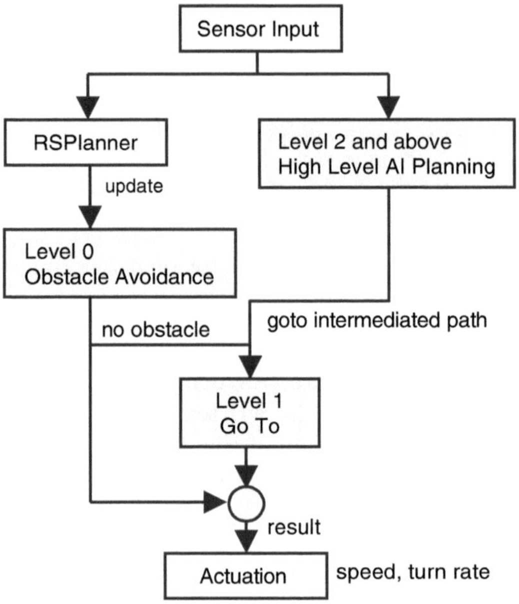

Fig. 11.6. Robotic simulator AI planning

11.4 AI Planner

In the RoboSim, there is an AI planner that provides the basic autonomous navigation capability to the simulated entities and serves as a foundation for higher level AI processing. Basic AI behaviours includes simple obstacle avoidance, path planning, waypoints following, leader following and road following. Figure 11.6 shows the hierarchy of the planner. In the planning process, lower level of planning implies higher priority. The simulator also provides interfaces for the option of adding plug-in for external AI control algorithm.

Table 11.1. Percentage error of time taken by a simulated robot to travel 100 m

Speed of simulated robot ($m\,s^{-1}$)	Average time taken by a simulated robot to travel 100 m				Theoretical time taken (s)	Percentage error (%)
	T_1 (s)	T_2 (s)	T_3 (s)	$T_{average}$ (s)		
0.2	496	501	504	500.33	500	0.06
0.5	201	200	200	200.33	200	0.16
1	102	101	100	101	100	1
2	50	51	51	50.67	50	1.33
	Average percentage error					0.6375

11.5 Testing of AI Algorithms

One of the objectives of developing RoboSim is to test different external AI algorithms for co-operative robotic tasks. Different operating scenarios requiring specialized algorithms can be tested. Simple algorithms like vehicle following, obstacle avoidances were tested to analyse the simulator performance. A third party AI algorithm for buried object detection, using multiple robots, was tested to study the client-server features as well as the robot co-operation behaviour. The results showed a good co-ordination between the robots and achieving comprehensive coverage of the search area.

Simple experiments were also carried out to analyse the physical realism – whether the physical performance of the robots in RoboSim is the same as, or close to, the performance of the robots in real time theoretically. Take for instance, the time taken for a robot to travel a distance of 100 m is (100/*speed of robot*) seconds, so in RoboSim, this operation is simulated and the time taken for the operation is observed and tabulated in Table 11.1. From Table 11.1, it is observed that the performance of RoboSim is near to the realism with a small percentage error of 0.6%. This would imply that the algorithm tested on RoboSim can be ported to the real robot with a higher confidence.

11.6 Real-Time Performance Test

Real-time performance is an important requirement for robotic simulator, which allows user to interact with the computer and have the sense of immersion. One factor that affects the sense of immersion, or the real-time performance, is the frame rate, or frame frequency, of the rendering of virtual environment. Sixty frames per second (fps), or Hertz is normally desirable to be interactive; and 15 fps is the lower limit for a real-time simulator where human need to visually track animated objects and react accordingly.

In RoboSim, the statistical data of the scene graph was obtained from the scene manager; and this includes the frame rates of the simulator. Figure 11.7

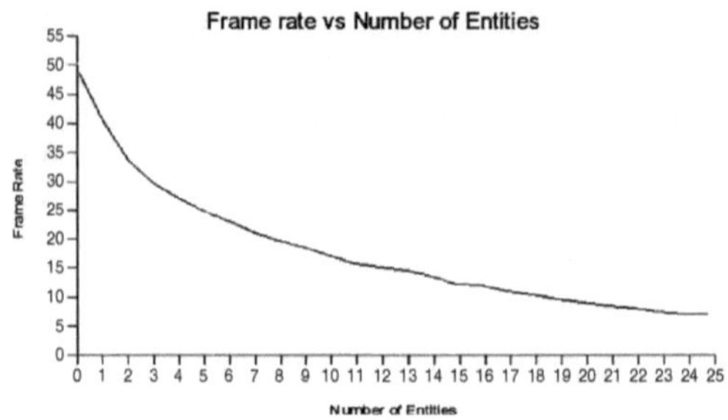

Fig. 11.7. Graph of frame rate vs. number of entities

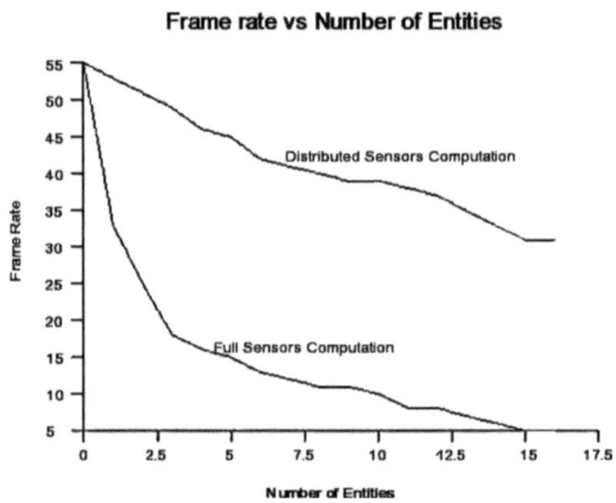

Fig. 11.8. Graph of frame rate vs. entity number with full and distributed sensors load

shows the result of frame rate vs. the number of entities running with full sensors loaded. It was observed that the frame rate dropped dramatically with the increased number of robot entities. This is mainly due to the heavy real-time computation for the sensors' readings and robots' collision model in every frame. Hence in Fig. 11.8, we can observe a less drastic frame rate drop when we employed the Network Collaborative Virtual Environment – simulation and rendering can be done on different machines connected through a network – using two computers over the Internet. As such, the computation is being distributed among the computers.

11.7 Conclusions

The design and development of a multi-mode 3D robotic simulator, for a PChost, was described. The multi-mode capability allows the use of a virtual environment and robots to test and verify advance AI algorithms. The simulator is further being developed as a platform to integrate both virtual and real robot, unifying the development of intelligent mobile robotics.

The RoboSim simulator combines the benefits of multi-mode configuration described and the use of simulators for fast effective evaluation of behaviour-based coded described, previously reported in [10, 11]. Simple experiments were conducted to verify the physical realism of RoboSim and real-time performance has been evaluated to conclude that having a NCVE would increase the simulation's performance of increasing number of entities. As such, RoboSim has been a useful tool to perform experiments and simulations to study behaviours of multiple autonomous agents.

References

1. J.S. Dahmann (1997), *High Level Architecture for Simulation*, 1st International Workshop on Distributed Interactive Simulation and Real-Time Application (DIS-RT'97), pp 9.
2. G. Seet, S.K. Sim, W.C. Pang, Z.X. Wu, T. Asokan (2006), *RoboSim: Multi-mode 3D Robotic Simulator For Study of Co-operative Robot Behaviour*, Proceedings of the Third International Conference on Autonomous Robots and Agents (ICARA 2006), pp 509–514, Palmerston North, New Zealand.
3. N. Koenig, A. Howard, *Gazebo - 3D multiple robot simulator with dynamics*, http://playerstage.sourceforge.net/gazebo/gazebo.html, visited on 1/9/2006.
4. O. Michel (2004), *WebotsTM: Professional Mobile Robot Simulation*, International Journal of Advanced Robotics Systems, Vol 1, No 1, pp 39–42.
5. D. Burns, *A New Processing Model for Multithreaded, Multidisplay Scene Graphs*, http://www.donburns.net/OSG/Articles/OSGMP/, visited on 28/3/2006.
6. P. Lindstrom, D. Koller, W. Ribarsky, L.F. Hodges, N. Faust, and G. Turner (1996), *Real-Time, Continuous Level of Detail Rendering of Height Fields*, Proceedings of SIGGRAPH 96, pp 109–118.
7. R.S. Johnston (1987), *The SIMNET Visual System*, Proceedings of the 9th ITEC Conference, pp 264–273, Washington D.C.
8. L.B. McDonald and C. Bouwens (1991), *Rationale Document for Protocol Data Units for Entity Information and Entity Interaction in a Distributed Interactive Simulation*, Institute for Simulation and Training Publication IST-PD-90-1-Revised, University of Central Florida, Orlando, FL.
9. E. Frcon and M. Stenius (1998), *DIVE: A scaleable network architecture for distributed virtual environments*, Distributed Systems Engineering Journal (Special Issue on Distributed Virtual Environments), Vol 5, No 3, pp 91–100.
10. X. Wang, G.G.L. Seet, M.W.S. Lau, E. Low, K.C. Tan (2000), *Exploiting Force Feedback in Pilot Training And Control Of An Underwater Robotics Vehicle: An Implementation In LabVIEW*, OCEANS 2000, MTS/IEEE Conference and Exhibition, Vol 3, pp 2037–2042.
11. J. Loh, J. Heng, G. Seet (1998), *Behaviour-Based Search Using Small Autonomous Mobile Robot Vehicles*, Second International Conference on Knowledge-Based Intelligent Electronic System, Vol 3, pp 294–301, Adelaide, Australia

12

A Mobile Robot for Autonomous Book Retrieval

Aneesh N. Chand and Godfrey C. Onwubolu

School of Engineering and Physics
University of the South Pacific, Private Mail Bag Suva
Fiji Islands

Summary. In this research we develop a robot intended to eliminate the human component in the book retrieval process by autonomously navigating to and retrieving a specific book from a bookshelf. Notably a salient feature of this autonomous robot is the employment of an elementary book detection and identification technique in the formulation of a time-optimal and computationally basic method in contrast to vision-based techniques that has been used by researchers in similar work.

Keywords: Autonomous book retrieval robot, book detection and identification.

12.1 Introduction

An emerging paradigm shift in robotic engineering developments is the development of service and task-oriented robots for providing assistance to humans in daily life. As such there are now available a broad genre of service robots including but not limited to, nursing and security robots, escort robots and hall polishing, cleaning and vacuuming robots [1, 2]. Such robots offer the merit of setting unprecedented levels of automation.

Commonly underlying to these robots is that they are developed for real-life applications that are usually labour-intensive and requires explicit human presence and effort; deployment of these robots then has the connotation of establishing a reduction in the need for human intervention in the task to be executed. This directly results in a reduction of human involvement and time and relegates humans to a safer but more potent role as planner. In addition, [3] cites numerous instances of autonomous robots being used in material handling and warehousing, in the assembly line and in the execution of common household chores.

In retrospect, we also designed a robot for a scenario that requires considerable human effort; the principal objective of this research is the design,

A.N. Chand and G.C. Onwubolu: *A Mobile Robot for Autonomous Book Retrieval*, Studies in Computational Intelligence (SCI) **76**, 101–107 (2007)
www.springerlink.com

formulation and synthesis of the hardware and software systems in the development of a Linux-based fully autonomous book retrieval robot. The current search for books is labour-intensive; this research is, therefore, aimed to develop a service robot that would automate the process of book searching in a typical library system. Moreover, emphasis was placed on the formulation of a computationally elementary but effective book identification method.

The operation of the developed system is as follows. A user searching for a book using the online database first retrieves the book call number. The user then enters the call numbers through a laptop computer onboard the robot. The robot navigates to the appropriate shelf and begins searching for the target book. Upon searching for the target book, the robot then retrieves the book and retraces its path back. Customized software written in a high level language (C++) provides the intelligence and custom made hardware provides the functionality for this robotic vehicle.

12.2 Hardware Design

The robot deployed for the purpose of book retrieval consists of a mobile wheel-driven platform base encompassing a gripper and book retrieval manipulation system with four axes of movements. The size of the robot is chosen such that it is able to accommodate a laptop computer (the principal controller) while simultaneously minimizing the weight, mechanical complexity and constructional cost. The maximum height of 1.5 m to which the gripper rises is attributed to the anticipated height of the bookshelves. This book retrieval robot is capable of retrieving a book of dimensions $(25 \times 17 \times 8)$ cm and of weight 1 kg.

A differential drive steering mechanism is used to provide locomotion to the robot with the main chassis being built over this steering base. The base has passive castors for stability and a line tracing mechanism to follow the navigation route. The navigation route is a white line leading to and running along a bookshelf. A two-finger gripper configuration is used to securely grip a book. The barcode scanner, gripper and book retrieval mechanism are constructed precisely over each other, as the barcode scanner provides the reference point for the book extraction execution. To provide vertical directionality to the barcode scanner, gripper and the book retrieval mechanism, a lead screw and pulley-based mechanism is used to provide vertical actuation so that the robot can scan and retrieve books from different heights with sensors indicating the different heights of the bookshelves. The control electronics encompasses direct current (DC) motor controller circuits, infrared sensors for obstacle detection, line tracer sensors for navigation, force sensors for the provision of force dexterity, limit switches and the respective interfacing circuits. The exhibit of Fig. 12.1 illustrates the book retrieval robot showing the major components.

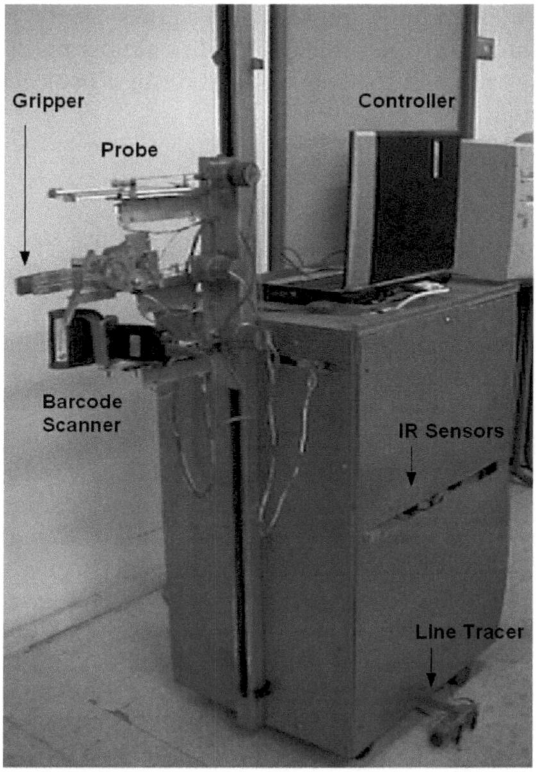

Fig. 12.1. The book retrieval robot

12.2.1 Book Retrieval Mechanism

The mechanism for systematically prying a book out of a shelf is constructed based on replication of the actions of the human hand. Based on these movements, the book retrieval mechanism was conceived and hence the *Probe* (the symbolic name allocated to the book retrieving mechanism) was constructed to replicate the human hand. This particular book retrieval mechanism was selected as it does not require the assumption that there are any spaces between the books. The *Probe* is a relatively simple mechanism made out of aluminium. It has inclined grooves fabricated, opposite to the direction of book retrieval motion, on the end that retrieves a book to create the necessary friction. A motor is used to actuate this back and forth horizontally on a linear slider with a nylon rope drive being used to transmit power from a motor shaft. Once the scanner has identified the target book and the robot has halted, the *Probe* extends horizontally into the bookshelf and is positioned precisely over the target book. The *Probe* then starts to move vertically down, applying a downward force on the book. When a certain magnitude of force (determined empirically) is attained, the *Probe* then retracts back horizontally. Since there is a sufficient downward force and friction also present, the book starts to tip on its end and into the gripper fingers. Additional details of the robot and a comprehensive overview of the robot operation in a book retrieval operation may be found in [4] and [5].

12.2.2 Electronics

The principal controller of the robot is a Pentium IV 1.5 GHz laptop computer onboard the robot. In addition, the control electronics of the book retrieval robot encompasses a number of electronic and transducer modules that are necessary for the operation of the book retrieval robot. All inputs to these circuits are transistor to transistor logic (TTL) levels. These are easily generated in software and transmitted through the parallel port.

Two rechargeable lead acid 12 V 7 A h^{-1} batteries have been used as the source of power. A simple regulator circuit regulates the 12 V supply to 5 V for the sensor and electronic circuits. With typical loading, the robot can continuously operate for approximately 90 min before recharging.

The onboard laptop requires an appropriate interfacing medium to physically interface with the control hardware. For this a custom designed, Enhanced Parallel Port (EPP) based, bidirectional, software selectable, input and output card encompassing a 74LS00 NAND gate, 74LS04 inventor, 74LS138 decoder, four 74LS373 tri-state latches, two 74LS244 buffers and various resistors for ensuring correct logic levels were used. All sensors, transducers and other electronic circuits are interfaced to either the input or output port of this parallel port interfacing card. All control signals are generated and transmitted through the output port and all sensory information from the sensors and transducers are interfaced to the input port of this card. The eight data pins (pins 2–9) of the total 25 pins of the parallel port are used for the transmission (as output) and reception (as input) of data and the selection of the interfacing card as either an input device or output device is implemented in software *on the fly* by selecting values of *strobe, data strobe* and *write enable* such that a particular 74LS138 decoder output is high that in turn activates the set of latches, two of which are used for outgoing data and the remaining two for incoming data. The value selection originates from the decimal value of each pin of the parallel port. A microcontroller board, acting as a secondary controller is also interfaced to the laptop through this interfacing card.

The secondary controller is a Microchip Technology PIC16F877 8-bit CMOS microcontroller with built in EPROM. This microcontroller is available as a 40-pin DIP package containing a central processor, EPROM, RAM, timer(s) and TTL/CMOS compatible or user defined input and output lines. This microcontroller is responsible for co-ordinating and receiving sensory information. Also, the interrupt feature of the microcontroller is heavily utilized for retroactive and instantaneous detection of sensor outputs.

The robot motion is controlled by the laptop parallel port using a 2-bit code sent to the motor controller circuits via the microcontroller board, via the parallel port interfacing card. The technology employed for motor controlling in this work was a motor diver monolithic integrated circuit (IC), specifically the SN754410NE quadruple half-H driver. This IC has two complete internal H-bridge circuits that can be used to independently power and control two separate DC motors and it is capable of supplying a continuous sustained current of 1 A and 2 A of current momentarily. The IC is available as a 16-pin DIP package containing internal protection diodes, anti sinking and sourcing circuitry and TTL and low-level CMOS compatible, high-impedance, diode clamped input lines.

Force feedback and dexterity is provided to the robot by means of a force sensing resistor (FSR). The Flexiforce A201 force sensor exhibits a decrease in resistance commensurate with an increase in the force subjected to its active sensing area and vice versa. Two of these force sensors have been utilized for force feedback in the book retrieval manoeuvres. The first sensor is attached to the underside of the *Probe* to determine the correct magnitude of force that suffices to retract a book out of the bookshelf while the second sensor, attached to the inside surface of one the gripper fingers determines the force with which the gripper securely grips a book. The magnitudes of these forces are determined empirically during the testing phase and then used as a reference set in the software. Quantification of the analog outputs of the force sensor to a discrete digital representation is executed using the internal 8-bit, self clocking and successive approximation analog to digital converter of the PIC16F877.

Sharp infrared sensors having a detection range of 80 ± 10 cm have been installed in the front and rear panels of the robot to detect any dynamic obstacles, should one obstruct the path of the robot. The interrupt handler for the event when an obstacle is detected generates and transmits logic low signals to the wheel motor controller circuits through the parallel port interfacing card so that for as long as an obstacle is present in the path, the robot halts momentarily. Finally, SLD-01 line tracer sensors are used for the robot navigation with 50 mm white lines leading to and running along the bookshelves representing the route the robot has to follow. A deviation of the robot locomotion to either side of the white lines causes the interrupt handler to execute correctional moves for the realignment of the robot with the navigation route.

A handheld PS2 port-interface type scanner with a scanning distance of 10 cm is used as the book identification device. Interfaced to the PS2 port of the laptop, the scanner scans the barcodes on the spines of books as the robots coasts by. Each book has a unique book call number digitally encoded in the barcode. The merit of using a barcode scanner is that, unlike vision-based techniques, the output data of the scanner is available digitally so that it may be processed unconditionally and instantaneously. Hence, the scanner scans the barcodes of books to match the scanned book call number with the user entered book call number. An affirmative match then results in a positive identification and the robot executes the book retrieval manoeuvres for the extraction of the subject book.

12.3 Software

The software that provides the robot cognition and intelligence is written in the C++ programming language under the Linux OS (Red Hat distribution version 9). Linux is chosen as it provides direct access to hardware and fares better than other conventional operating systems such as Windows in real-time hardware operations.

All behaviours of the robot (mechanical actions) have been represented as functions and all attributes are represented as data variables in the software. Invocations of these functions in their required sequences attain the book retrieval manoeuvres. Logically related functions and data are then grouped together and have been implemented in classes as methods and attributes that imply object oriented (OO) methodology. The software development phase has utilized the OO methodology to attain abstraction and modularity.

The software developed for the barcode scanner is fundamentally a half duplex, interrupt driven, software implementation of PS2 communications. Reception of a byte written at the PS2 port by the barcode scanner is automatically invoked whenever the barcode scanner scans in a barcode. The software codes for the data retrieval are part of standard C++ language/libraries and hence guarantees that the code used to retrieve data from the PS2 port will be platform independent and thus compile under any platform while simultaneously having the merit of being machine independent. Since the software retrieval of data scanned by the barcode scanner uses elementary C++ programming, it does not constitute computation overhead as is the case with using computer vision for book detection and identification as done by [6–8].

12.4 Experimental Results

Several experiments were carried out for the validation and verification of the performance of the system. The major metrics of measurement were the optimum robot *speed* that suffices to accurately scan book barcodes, the *accuracy* of and the *time* required for the detection and identification of books. The accuracy and book seek-time analysis is important in comparing the results of other researchers and to determine the efficacy of the barcode scanner as the book identification device. From current experimental results, the robot is capable of correctly detecting and identifying books with 99% accuracy at a speed of $0.6\,\mathrm{m\,s^{-1}}$. It takes an average time of 0.45 s for the robot to detect, scan and realize the identity of a single book and is able to successfully retrieve books 53% of the time. These results were obtained in a controlled environment encompassing a number of required assumptions such as lighting conditions, presence of obstacles and dimensions of books.

12.5 Conclusion

The design, development and deployment of a prototype book search and retrieval robot has been presented and successfully attained. Customized cross-platform compatible software, written in C++ language, forms the intelligence of the robot while an onboard laptop computer, running on Linux platform of Red Hat distribution version 9.0, provides the processing and execution of intelligence of the robot. A Graphical user interface (GUI) developed using Borland's Kylix® acts as the human–machine interface allowing a user to submit a desired book call number. DC motors are used for locomotion. Infrared line tracer sensors are used for the navigation of the robot with lines embedded in the floors marking the routes and giving the robot a priori knowledge of the paths it has to take. Barcoding technology is used to provide digital signatures to library books, encoding its call number into the barcode. The employment of an innovative and practical system, a typical and inexpensive barcode reader, provides the technology with which books are searched for. Current ongoing work includes improvising the performance of the robot and incorporation of features such as circumnavigation around obstacles, localization and automatic self-recharging that will solve the problems of deployment in a realistic environment.

References

1. Carnegie DA, Loughnane DL, Hurd SA (2004) The design of a mobile autonomous robot for indoor security applications. J Eng Man 218: 533–543
2. Ohya A, Nagumo Y, Gibo Y (2002) Intelligent escort robot moving together with human-methods for human position recognition. In: Joint 1st Int Conf on Soft Compyting and Intelligent Systems and 3rd Int Symposium on Advanced Systems. Published in CD-ROM
3. Fukuda T, Hasegawa Y (2003) Perspectives on intelligent control of robotics and mechatronic systems. In: 2nd Int Conference on Computational Intelligence, Robotics and Autonomous Systems, Published in CD-ROM
4. Chand AN (2006) Development of a book retrieval robot. MSc Thesis, University of the South Pacific
5. Chand AN, Onwubolu GC (2006) Automatic book detection and retrieval using an autonomous robot. In: Mukhopadhyay SC, Gupta GS (eds) Third Intl Conf on Autonomous Robots and Agents, Palmerston North, pp 15–20
6. Prats M, Ramos-Garijo R, Sanz PJ, Pobil AP (2004) An autonomous assistant robot for book manipulation in a library. In: Intelligent Autonomous Systems 8, IOS Press, pp 1138–1145
7. Tomizawa T, Ohya A, Yuta S (2003) Remote book browsing system using a mobile manipulator. In: IEEE Intl Conf on Robotics and Automation, pp 3589–3594
8. Tomizawa T, Ohya A, Yuta S (2004) Book extraction for the remote book browsing robot. J Rob and Mech 16: 264–270

13

Trajectory Planning for Surveillance Missions of Mobile Robots

Luiz S. Martins-Filho[1] and Elbert E.N. Macau[2]

[1] Department of Computer Science, Federal University of Ouro Preto, Ouro Preto, MG, Brazil luizm@iceb.ufop.br

[2] Associate Lab for Computing and Applied Mathematics, National Institute for Space Research, S. José dos Campos, SP, Brazil elbert@lac.inpe.br

Summary. This paper discusses a trajectory planning strategy for surveillance mobile robots. This specific kind of mission requires fast scanning of the robot workspace using trajectories resembling non-planned motion for external observers. The proposed trajectory planner is based on the typical behaviour of chaotic dynamical systems. A chaotic behaviour is imparted to the robot using a trajectory-planner based on the well-known Standard map.

Keywords: Mobile robots, motion planning, surveillance robots, kinematic control, chaos.

13.1 Introduction

This work addresses issues on mobile robots motion planning and control to accomplish a specific class of missions: the surveillance of terrains for security purposes [1]. These surveillance missions require a motion control system comprising some basic navigation competencies, like planning and reacting. Moreover, if we consider eventual intruders with intelligent capabilities to try to avoid the sentry robot, additional motion requirements are strongly desirable: high unpredictability of the robot trajectories and fast workspace scanning. To satisfy these special demands, we propose a trajectory planner based on dynamical behaviour of chaotic systems, working as an auxiliary module within a closed-loop locomotion control scheme. The motion planning results on a trajectory passing through a planned sequence of target locations.

Studies concerning the interactions between mobile robots and chaos have been developed more recently. An integration between the robot motion and the Arnold chaotic system, for instance, is proposed to impart chaotic behaviour to mobile robots in [2]. In [3], an extension of this same strategy, applying different chaotic systems on integration with the robot kinematic, is presented.

L.S. Martins-Filho and E.E.N. Macau: *Trajectory Planning for Surveillance Missions of Mobile Robots*, Studies in Computational Intelligence (SCI) **76**, 109–117 (2007)
www.springerlink.com

In [4], an open-loop control approach is proposed to produce unpredictable trajectories, using Lorenz chaotic system to command the wheels.

An important issue on surveillance missions for robots is the patrolling path planning. Actually, the motion strategy is crucial for the success of any surveillance mission. For accomplishing the requirement of complete terrain scanning, an intuitive strategy is to command the robot to cross every region, covering systematically the entire terrain. A quite obvious solution is a systematic scan using parallel straight trajectories. Nevertheless, a smart intruder could easily understand this strategy. Using a group of robots, the patrolling planning becomes an easier problem, where each robot receives a scan task for a terrain sub-area, optimizing the actions [5]. In [6], another approach is proposed, applying genetic algorithms to maximize the terrain scan in an intricate strategy, and with high computational costs. A traditional approach is based on human operators commanding the robots to visit regions in an arbitrary way, or in basis of suspections about the intruder presence [5]. In terms of unpredictability, the trajectories produced using random motion can be very erratic and impossible to be understood by the intruder [7].

In this work, we intend to exploit the dynamical behaviour of a conservative chaotic system to obtain adequate planned trajectories to be followed by the sentry robot, accomplishing the main surveillance mission requirements.

13.2 Trajectory Planning

Our objective is to obtain a deterministic chaotic behaviour of the robot motion, in such a way that the trajectories become unpredictable, i.e. the knowledge of the robot system state during an arbitrary time does not allow to predict its subsequent trajectory. That is a consequence of the initial conditions dependence or sensibility, and the continuous frequencies spectrum, that characterizes a non-periodic behaviour. This behaviour is an attractive issue for the considered mobile robots missions since it provides the desired trajectory unpredictability and fast terrain scanning, and it can be imparted to the robot using Poincaré sections. A Poincaré section defines a dynamic system's phase space with arbitrary dimension, and composes a map featuring the chaotic behaviour. Differential equations generate a discrete map through the flow intercepting a Poincaré section at each unity of time.

A well-known two-dimensional coupled return map in phase space, called Standard map, represents a physical system defining a kicked rotor, and is very useful for studying the chaotic motion basic features. This area-preserving map was firstly proposed by Bryan Taylor (and independently obtained by Boris Chirikov [8]) to describe the dynamics of magnetic field lines on the kicked rotor [9]. The experimental scheme of this system is composed by a rotor subjected to a driving force that comes at exactly even intervals in time, provided by a periodic pulse. The simple scheme consists of kick occurrence on a constant direction and intensity. The consequence of this forcing action on the systems dynamics depends of the system position and speed at the kick application instant. The mathematical model of a Poincaré section of the phase portrait of this system is given by the following map equations

$$\begin{cases} x_{i+1} = x_i + K \sin y_i \\ y_{i+1} = y_i + x_{i+1} \end{cases} \tag{13.1}$$

Here x is a periodic configuration variable (angular position) and y is the momentum variable (angular speed), both computed mod(2π). The map is parameterized by K, the strength of the non-linear kick. Standard map is so interesting mainly because it can show chaos under very basic conditions. The terrain coverage using the Standard map is analysed by iterating the map and verifying if the sequence of planned positions satisfies the requirements. We define a 100×100 (normalized measurement unit) square terrain. The sequence begins with an arbitrary initial position, and considers $K = 7$. The result for 10,000 iterations is shown in Fig. 13.1, where the uniform distribution of the points over the whole terrain can be observed (necessary condition for complete scan is $K > 6$). We define a performance index for the terrain coverage based on square unit cells (1×1), and we analyse the visited cells percentage for the locations planning (we do not consider the robot trajectories between two target locations). This visitation index, plotted in terms of its time evolution, is shown in Fig. 13.2 (index=1 represents 100% of cells).

These results are similar to a uniform distribution obtained by random numbers generation, however, with different construction nature, as we discuss in Sect. 13.4. The time evolution of visitation index allows us to conclude that the complete scan is ensured. Moreover, the robot will also visit cells during the displacement between two consecutive planed positions.

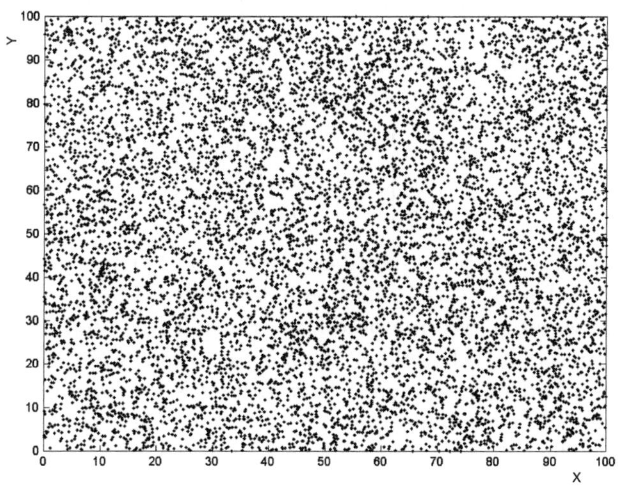

Fig. 13.1. The terrain showing the subgoals locations planned by the Standard map

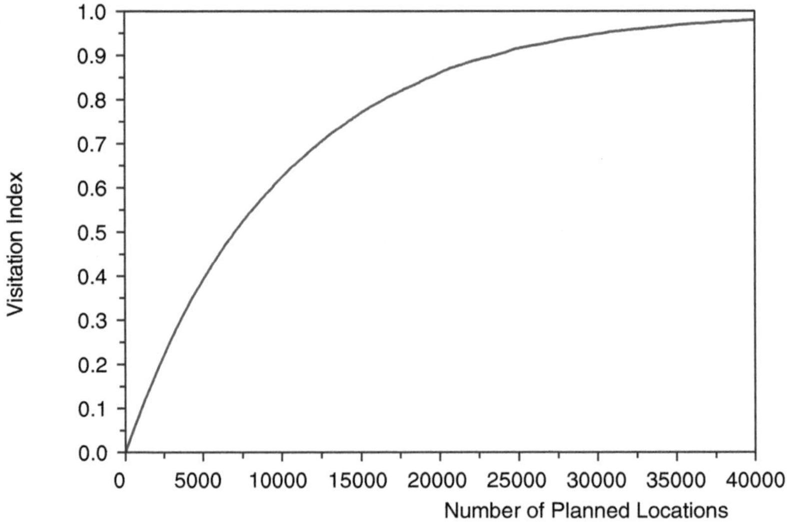

Fig. 13.2. Time evolution of visited cells index (40,000 iterations)

13.3 Robot Motion Control

We adopt here a differential motion robot with two degrees-of-freedom. The locomotion system is composed by two independent active wheels and a passive wheel working as a free steered wheel. The robot is equipped with proximity sensors capable of obstacles detection providing short-range distances, and with specialized sensors for detection of intruders. The robot body is considered rigid, remaining on the horizontal plane. The motion resulting of a kinematic control of the active wheels can be described in terms of robot position and direction ($x(t)$, $y(t)$, $\theta(t)$), and linear and angular velocities ($v(t)$, $\omega(t)$). The motion control is done by providing the two active wheels velocities or $v(t)$ and $\omega(t)$. The mathematical model of this motion is given by [10]:

$$\begin{bmatrix} \dot{x} \\ \dot{y} \\ \dot{\theta} \end{bmatrix} = \begin{bmatrix} \cos\theta & 0 \\ \sin\theta & 0 \\ 0 & 1 \end{bmatrix} \begin{bmatrix} v \\ \omega \end{bmatrix} \tag{13.2}$$

This equation describes a non-holonomic non-linear system. The control problem for this class of systems has been deeply studied, and diverse adequate solutions are available. We adopted one solution involving a state feedback controller proposed in [11], which is an appropriate approach to produce a trajectory described by a sequence of coordinates (x_p, y_p). It means that the target passage locations planning is performed by a specialized robot module, independent of the robot motion control module. In other words, the planning module provides the locations sequence to the control module.

This control law considers the geometric configuration described in Fig. 13.3. The robot is posed at an arbitrary position and orientation (x, y, θ),

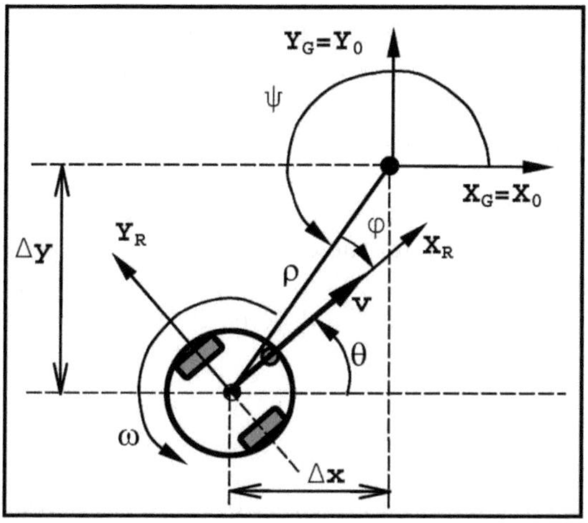

Fig. 13.3. Configuration of the kinematic control showing the robot position x, y and orientation θ, the error variables ρ and φ, and the control variables v, ω

and the target position is defined by the trajectory-planner. The robot kinematic model (13.2) is described on the workspace absolute reference frame. We define φ as the angle between the direction from the robot centre to the desired position and the axis X_R of the body reference frame. The variable ρ describes the distance between current and desired positions, and ψ is the angle between the direction to the desired position and X_0. We simplify the model, reducing the problem to a two-degrees-of-freedom one, using polar coordinates and a coordinate change: $\rho = (\Delta x^2 + \Delta y^2)^{1/2}$ and $\varphi = \pi + \theta - \psi$. The new mathematical description is given by

$$\begin{bmatrix} \dot{\rho} \\ \dot{\varphi} \end{bmatrix} \begin{bmatrix} -\cos\varphi & 0 \\ \frac{1}{\rho}\sin\varphi & 1 \end{bmatrix} \begin{bmatrix} v \\ \omega \end{bmatrix} \tag{13.3}$$

This new motion model is not defined at $x = y = 0$, i.e. at the origin of $X_G Y_G$ frame ($1/\rho \to \infty$ for $\rho \to 0$). This point is only achieved when the robot reaches the goal location, and it can be avoided switching the target to the next planned location before arriving too close to the singularity. The feedback control law, defining the system inputs v and ω, is given by $v = k_1 \rho \cos\varphi$ and $\omega = -k_1 \sin\varphi \cos\varphi - k_2 \varphi$. The proof of the system stabilization, i.e. the convergence to the target position, is given by Lyapunov stability analysis. Considering a Lyapunov function V defined in terms of ρ and φ, $V = (1/2)(\rho^2 + \varphi^2)$, taking its derivative, and using (13.3), we obtain

$$\begin{aligned} \dot{V} &= \rho(-v\cos\varphi) + \varphi(\omega + \tfrac{v}{\rho}\sin\varphi) \\ &= -k_1(\rho\cos\varphi)^2 - k_2\varphi^2 \end{aligned} \tag{13.4}$$

Considering exclusively positive values for k_1 and k_2, we can see that $\dot{V} \leq 0$ $\forall(\rho, \varphi)$, a sufficient condition for the asymptotic convergence of (13.3).

A second control approach, called here discontinuous control law, is defined as a sequence of two control steps: an initial rotational maneuver around the robot centre to orientate the robot head towards the goal position, followed by a straight trajectory motion directly to the planned objective point.

Using these two control laws, we test the proposed planning strategy, and analyse the unpredictability and the coverage of the robot trajectories. We do not consider in this work the problem of obstacle avoidance; nevertheless, a simple solution for this problem could be easily implemented.

13.4 Simulations Results

The robot motion was simulated numerically using the mobile robot kinematic model discussed above, applying a motion control to track the trajectory composed of a planned sequence of objective points. In the case of different terrain shapes, the planning process could fit the interest area inside a square Standard map, excluding the points planned outside the terrain. Moreover, the robot can percept an object or event when performing the surveillance task inside the sensor range region. The dimensions of this perception field depend on the properties of the device used to perceive external objects. In this work, we do not take into account this extra area covered by the sensors.

Considering the crucial requirements of a typical surveillance mission, and the main ideas developed here, we can configure different scenarios for the simulations, combining motion control laws and planning approaches. We compose here two scenarios. In the first case, we consider the adopted continuous control law that can produce smooth trajectories avoiding undesirable manoeuvres and control switches, and we apply the planning based on the Standard map to establish a locations sequence to be visited. An example of trajectory obtained by numerical simulations for this scenario can be seen in Fig. 13.4.

In the second scenario, a random strategy is applied to the trajectory planning, taking a random sequence of locations uniformly distributed in the patrolled area. This strategy provides a result similar to the chaotic one (cf. Sect. 13.2). For the motion control, the discontinuous control law is adopted: rotational maneuver followed by a straight trajectory. The results for an example of this case are shown in Fig. 13.5.

Examining the options of control laws, the continuous one offers advantages: smoother trajectories reduce control switches and manoeuvres, contributing to perform unpredictable trajectories for external observers, and reinforcing the erratic aspect. In the opposite case, the discontinuous control composed of two subsequent manoeuvres (a robot rotation around itself followed by a linear displacement directly to the desired location) produces piece-wise straight trajectories that could be predicted by an external observer.

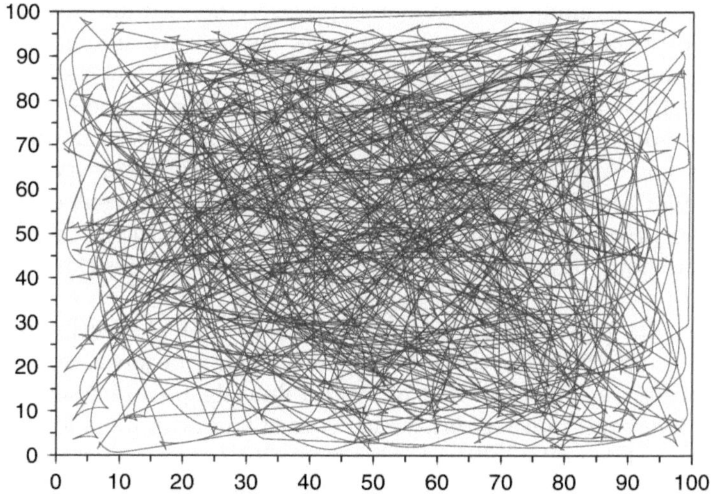

Fig. 13.4. An example of the mobile robot trajectory evolution considering the first case: combining chaotic path-planning and continuous motion control

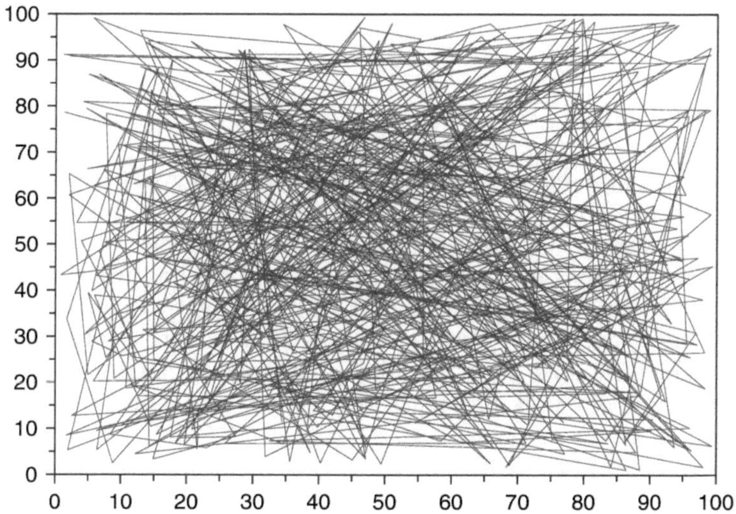

Fig. 13.5. An example of the mobile robot trajectory evolution considering the second case: combining random path-planning and discontinuous motion control

The choice between chaotic and random planning can result in similar results in terms of surveillance space coverage. However, there is a fundamental difference between chaotic and random trajectories: the deterministic nature of the sequence of locations. Considering the chaotic planning approach, the robot navigation system maintains complete control and knowledge of the planned passage positions sequence for the reason that it is a deterministic system and its dynamical behaviour is precisely defined and deeply studied. In terms of basic navigation competencies, this path-planning determinism represents an important advantage over navigation based on a sort of random walk trajectory. This determinism can facilitate the frequent robot localisation procedure, which is a crucial function because the knowledge of the robot position with appropriate precision is a very necessary information for the robot itself and also for the mission operation centre. In a robot localisation procedure, sensors data assimilation algorithms take advantage of good previous information to estimate the current position with better accuracy.

The results presented in Sect. 13.2 confirm that the entire terrain will be visited and verified by the robot. When moving between two subsequent planned positions, the robot visit other terrain cells, possibly passing multiple times by diverse terrain regions, which consists an advantage for the mission of search a mobile and intelligent intruder trying to scape of the sentry detection. The erratic aspect of the robot motion can be easily verified by visual inspection of the obtained trajectories that allows us to conclude qualitatively that the main mission requirements (trajectories unpredictability and complete terrain scan) are supplied by the proposed chaotic path-planning strategy.

13.5 Conclusion

The strategy proposed in this work can obtain appropriate results suited for the path-planning of surveillance mobile robots. The planned sequence of objective passage locations on the trajectory satisfies the main requirements of fast scanning of the robot workspace. The resulting chaotic motion behaviour ensures high unpredictability of the robot trajectories, resembling a non-planned and quite erratic motion from external observers' point of view. The proposed path-planning strategy has an important advantage over a random walk-like path that is the deterministic nature of the planning.

Moreover, this work makes us believe that applications of special dynamical behaviours of non-linear systems on solutions for robots motion problems represent an attractive interdisciplinary interface for researchers of both scientific domains, reinforcing the interesting perspectives of future works.

Acknowledgements

The authors acknowledge the financial support of Cons. Nac. de Desenvolvimento Científico e Tecnológico - CNPq, Fund. de Amparo à Pesq. de M. Gerais - FAPEMIG, and Fund. de Amparo à Pesq. de S. Paulo - FAPESP (Brazil).

References

1. Martins-Filho LS, Macau EEN, Rocha R (2006) Planning of unpredictable trajectories for surveillance mobile robots. In: Proc Int Conf Autonomous Robots and Agents. Massey Univ, Palmerston North
2. Nakamura Y and Sekiguchi A (2001) IEEE Trans Robotics and Automation 17(6): 898–904
3. Jansri A, et al. (2004) On comparison of attractors for chaotic mobile robots In: Proc Conf IEEE Indus Elect Soc. IEEE, Busan
4. Martins-Filho LS, et al. (2004) Commanding mobile robots with chaos. In: Adamowski JC, et al. (Eds) ABCM Symposium Series Vol. I. ABCM, S. Paulo
5. Feddema J, et al. (1999) Control of multiple robotic sentry vehicles. In: Gerhart GR, et al. (Eds) Proc. of SPIE - Unmanned Ground Vehicle Tech Vol. 3693. SPIE, Orlando
6. Emmanuel T, et al. (1994) Motion planning for a patrol of outdoor mobile robots. In: Proc Euro Rob and Intel Sys Conf. Malaga
7. Heath-Pastore T, et al. (1999) Mobile robots for outdoor security applications. In: Proc Int Topical Meeting Robotics and Remote Sys. ANS, Pittsburgh
8. Chirikov BV (1979) Physics Reports 52(5): 263–379
9. Lichtenberg AJ, Lieberman MA (1983) Regular and Stochastic Motion. Springer-Verlag, Berlin
10. Siegwart R, Nourbakhsh IR (2004) Introduction to Autonomous Mobile Robots. MIT Press, Cambridge
11. Lee SO, et al. (2000) A stable target-tracking control for unicycle mobile robots. In: Proc IEEE/RSJ Int Conf Intel Rob and Sys. IEEE, Takamatsu

14

An Evolutionary Approach to Crowd Simulation

Tsai-Yen Li and Chih-Chien Wang

Computer Science Department
National Chengchi University
li@nccu.edu.tw, g9022@cs.nccu.edu.tw

Summary. In previous work, virtual force has been used to simulate the motions of virtual creatures, such as birds or fish, in a crowd. However, how to set up the virtual forces to achieve desired effects remains empirical. In this work, we propose to use a genetic algorithm to generate an optimal set of weighting parameters for composing virtual forces according to the given environment and desired movement behaviour. A list of measures for composing the fitness function is proposed. We have conducted experiments in simulation for several environments and behaviours, and the results show that compelling examples can be generated with the parameters found automatically in this approach.

Keywords: Crowd simulation, Genetic algorithm, Robot formation, Multiple robot system.

14.1 Introduction

Formation control for multi-robot systems is a classical robotic problem that has attracted much attention in the literature [1]. In recent years, the techniques of simulating virtual crowd also have created potential applications in contexts such as a virtual mall, digital entertainment, transportation and urban planning. Most of the current systems for crowd simulation adopt a local control approach. One of the common approaches to this problem in computer animation adopts the virtual force model [10], whereby the movement of each robot is affected by virtual forces computed according to its spatial relation between itself and other robots or objects in the environment. However, designers still need to face the problem of how to choose the most appropriate forces and select the best weights. In addition, there is no objective way to evaluate the result.

Crowd motions can be generated by simulation or by planning. In this work, we focus on the approach of simulation. Bouvier [2], Brogan and Hodgins [3] used a physical-based particle system to simulate a crowd of athletes such as

T.-Y. Li and C.-C. Wang: *An Evolutionary Approach to Crowd Simulation*, Studies in Computational Intelligence (SCI) **76**, 119–126 (2007)
www.springerlink.com

runners or bikers competing in a field. Reynold [10,11] sought to apply virtual forces to individual robots to create steering behaviours for the whole crowd. Tu and Terzopoulos [4] used rule-based finite-state machine to construct a cognitive model of fish and succeeded in creating several interesting behaviours including flocking. Muse and Thalmann [9] used behaviour rules to design virtual characters and used scripts to construct complex social interactions.

In this chapter, we propose to model the problem of generating good examples of the movement of crowds by designing an appropriate parameterization and evaluation mode and adopt the genetic algorithm [7] to search for an optimal set of parameters. The parameterized virtual-force model is then evaluated by an array of measures to describe the desired crowd behaviours. We have also conducted experiments to generate parameter sets for several common crowd behaviours in various environments consisting of different spatial structures. The considered materials are derived from [6] where they were first presented and discussed.

14.2 Design of Movement Model and Virtual Forces

In this work, we have adopted the virtual force model proposed in [10] as the way to affect the movement of each individual robot. We assume that the robots move under the influence of virtual forces proposed in [11], and that they must respect maximal speed limits in both translation and rotation. In addition, a robot is given a view angle of 330 and a constant view distance that is 20 times the size of the robot. Because the virtual forces are computed locally for each robot, only the robots or obstacles that are within the range of view have effect on the computation of the virtual forces for the robot. The computed forces are used to update the next configuration of the robot. However, if the new configuration is in-collision, the system makes use of various local strategies to avoid such a collision.

The motions of the crowd in our simulation system are driven by five types of virtual forces: *separation, alignment, cohesion, following* and *collision*. First, the separation force (F_{sep}) is a repulsive force computed proportional to the distance between the robot and other neighbouring robots within the range of view. The resultant force is the summation of all of the forces exerted by each individual robot. The effect of the force is to maintain a safe distance between the robots. Second, the alignment force (F_{align}) is used by a robot to align its velocity and orientation with other neighbouring robots. An average velocity for the robots within the range of view is computed first. The alignment force is then computed according to how the velocity of the current robot deviates from the average one. Third, the cohesion force (F_{coh}) is computed according to the difference vector between the current position of the robot and the centre of the neighbouring robots within the range of view. Fourth, the following force (F_{fol}) is an attractive force that drives the crowd to its goal. This force is computed according to the distance from the goal, which could also be moving. We have adopted the model proposed in [5] to compute a collision-free following path by making the robot head to a point along the trace of the leader that does not cause collisions with obstacles. Finally, the collision force (F_{col}) is a repulsive force exerted by the environmental obstacles when a condition of collision is predicted after certain period of time.

Our system uses a linear combination of the normalized component forces described above to compute the final virtual force as shown in (14.1).

$$F = s_1 * F_{sep} + s_2 * F_{align} + s_3 * F_{coh} + s_4 * F_{fol} + s_5 * F_{col} \tag{14.1}$$

The set of weights $s = (s_1,\ s_2,\ s_3,\ s_4,\ s_5)$ determines how the forces are composed to affect the behaviour of the robot. The appropriate setting of the weights remains as an empirical problem for the designer, which is the reason that we propose to automate the search process for an optimal solution with the genetic algorithm described in the next section.

14.3 Evolution of Motion Behaviour of Crowd

In a crowd simulation, the trajectory of the crowd may vary greatly according to the scenario and the environment in which the crowd is situated. Therefore, the designer is required to tune the weights of the component forces in order to achieve the results desired within the framework of virtual forces. The complexity of the parameter space and the time-consuming process of developing the simulation make it a difficult optimization problem for human or for machine. Therefore, we propose to solve the problem by using a genetic algorithm, commonly used to solve the problem of searching for an optimal solution in a large search space [7].

In our problem, the set of weights described in the previous section are used as the genes for encoding. Each of genes is encoded into a bit string of length 10. Since we have five parameters (genes) in our system, the total length of the chromosome is 50 bits. In the current system, the population is set to 200. We have chosen the roulette wheel selection mechanism to select the samples that will survive in the next generation. Sample points are randomly selected in the wheel for the one-point crossover operation. In each generation, we also perform the mutation operation (switching a random bit) on samples selected with the probability of 0.01.

The desired fitness function in our experiments may also be different for different scenes and different types of behaviours. Instead of designing a specific fitness function for each type of behaviour, we designed several elementary fitness functions that can be used to compose the final fitness function of a behaviour. Most of the elementary fitness functions are computed based on the spatial relation between a robot and its neighbours, defined as the k-nearest robots. Following Miller [8], we currently set the value of k as seven in the current system. We describe the elementary fitness functions used in this work as follows.

Inter-robot distance (G_m). The inter-robot distance for a robot is defined as the average distance of its k-nearest neighbours. The relative distance to robot i is denoted as r_i, and their average distance between each other is denoted as R_i as shown in (14.2). For a given frame, the difference between the average and the user-specified value is the main performance index. This value is normalized by the quantization factor Q_d to make it fall in the interval $I_f = [0, 1]$, as shown in (14.3). The overall system performance is computed as the average of all robots over the whole path as shown in (14.4).

$$R_j = \frac{\sum_{i=1}^{i=k} r_i}{k} \tag{14.2}$$

$$G_j = \frac{Q_d}{|R_j - R_e|}, \quad \text{s.t. } 0 < G_j \leq 1 \tag{14.3}$$

$$R_m = \frac{\sum_{j=1}^{j=N} G_j}{N}, \quad G_m = \frac{\sum_{m=1}^{m=L} R_m}{L} \tag{14.4}$$

Distance to the goal (G_g). The calculation is similar to that for inter-robot distance described above except for that the distance is computed with respect to the goal instead of each robot. In the interest of saving space, we do not repeat the formula here. In addition, instead of being a physically existing object, the goal could also be a designated position at a point behind a possibly moving leader.

Number of collisions (G_c). Assume that N_c denotes the number of robots that are in collision with other robots or obstacles, and that N denotes the total number of robots in the simulation. Then, the collision ratio s_j is defined as the percentage of robots that are in collision, and the overall elementary fitness function is then defined as the average of this ratio in movement over the whole path as shown in (14.5).

$$s_j = \frac{N - N_c}{N}, \quad G_c = \frac{\sum_{j=1}^{j=L} s_j}{L} \tag{14.5}$$

Consistency in orientation (G_a). In (14.6), we define the average difference in the orientation of a robot with respect to other neighbouring robots and the consistency in orientation for a single robot. The fitness function for the whole crowd is then defined as the average of all robots over the whole path as shown in (14.7).

$$A_j = \frac{\sum_{i=1}^{i=k} |\theta_j - \theta_i|}{k}, i \neq j \qquad B_j = (1 - \frac{\mathrm{A}_j}{\pi}) \tag{14.6}$$

$$B_m = \frac{\sum_{j=1}^{j=N} B_j}{N}, G_a = \frac{\sum_{m=1}^{m=L} B_m}{L} \tag{14.7}$$

Consistency in distance (G_d). The average distance and standard deviation of a robot's distances to its neighbours are first computed according to (14.8).

The consistency in distance for a robot is then defined as the inverse of the standard deviation multiplied by the quantization factor Q_σ in (14.9). The overall performance index is computed as the average of the consistency in differences between all robots over the whole path as shown in (14.10).

$$R_{mean} = \frac{\sum_{i=1}^{i=k} r_i}{k}, \sigma_j = \frac{\sqrt{\sum_{i=1}^{i=k} (R_i - R_{mean})^2}}{k} \tag{14.8}$$

$$F_j = \frac{Q_\sigma}{\sigma_j}, \text{ s.t. } 0 < F_j \leq 1 \tag{14.9}$$

$$F_t = \frac{\sum_{j=1}^{j=N} F_j}{N}, G_d = \frac{\sum_{t=1}^{t=L} F_t}{L} \tag{14.10}$$

The overall fitness function (F_{sum}), as shown in (14.11), is computed based on a linear combination of the elementary fitness functions defined above.

$$G_{\text{sum}} = S_m * G_m + S_g * G_{\text{g}} + S_c * G_{\text{c}} + S_a * G_a + S_d * G_d \tag{14.11}$$

A designer makes use of the weights, $S = (S_m, S_g, S_c, S_a, S_d)$, according to the nature of the desired behaviour, to compose the final fitness function from the elementary ones. We assume that these weights are more intuitive to set compared to the weights in (1) and that they should be the same for the same desired behaviour of crowd movement. However, the optimal weights in (1) may be scene specific, as shown in the next section.

14.4 Experimental Design and Results

We implemented the simulation system and the genetic algorithm in Java. The evolution terminates when the maximal number of generations is reached or the optimal solution converges and do not change for five generations. For each environment and behaviour, we ran the same experiment with three different initial settings to see if they all converge to the same optimal solution. Our observation is that most of the experiments converged after 17–19 generations.

Generally speaking, it is difficult to classify environments or define typical scenes. Nevertheless, we have defined and tested three types of scenes that we regard as typical in our current experiments. These scenes include an open space without obstacles (E1), a scene with a narrow passage (E2) and a scene cluttered with small obstacles (E3).

Three types of behaviours were tested in our experiments: *group moving* (B1), *following* (B2) and *guarding* (B3). The group moving behaviour refers to keeping the crowd moving with a given inter-robot distance in a group. The following behaviour refers to pursuing a possibly moving goal as closely as possible, and the guarding behaviour refers to surrounding a possibly moving goal. For each different type of behaviours, we used a different set of elementary fitness functions to compose the final fitness function in order to express the designer's intention.

We conducted experiments with the genetic algorithm described in the previous section to acquire the set of parameters for the desired behaviour for each given environment. The set of weighting parameters generated by the system are given in Table 14.1. These optimal parameter sets vary greatly for different scenes and for different types of environment. For example, for the behaviour B1, the parameter s_3 varies from 0.15 to 0.99 for different environments. For the environment E1, the parameter s_3 varies from 0.19 to 0.95. The generated weights for the following force (s_4) and the collision forces (s_5) for the cluttered environment (E3) are relatively smaller than other environments since these two forces tend to jam-pack the crowd in such an environment.

In order to validate the parameters, we used the parameters obtained for E1 to run simulations with E2 and E3. The overall scores G_{sum} returned by the fitness function are 295, 150 and 262, respectively. When we used the optimal parameters generated for E2 and E3, respectively, to run the experiments again, the scores improved to 271 and 284, respectively. Although the scores in the cluttered environments (E2 and E3) are not as good as might have been anticipated from the result for E1, the scores were greatly improved when the optimal parameters were used. This experiment reveals that the optimal weighting parameters for the virtual force are scene dependent.

In Fig. 14.1, we show an example of the following behaviour (B2) for the crowd with a small inter-robot distance in a space cluttered with obstacles (E3). In Fig. 14.2, we show an example of group movement (B1) where the desired inter-robot distance is set to a higher value, and the crowd needs to pass the narrow passage (E2) in order to reach the goal.

Table 14.1. Optimal weights generated by the genetic algorithm for various environments and behaviours (B: Behaviour, E: Environment)

	s_1	s_2	s_3	s_4	s_5
B1-E1	0.42	0.03	0.15	0.84	0.44
B1-E2	0.87	0.31	0.99	0.49	0.16
B1-E3	0.65	0.12	0.19	0.01	0.12
B2-E1	0.53	0.15	0.77	0.36	0.62
B2-E2	0.90	0.39	0.80	0.06	0.16
B2-E3	0.95	0.21	0.93	0.07	0.04
B3-E1	0.94	0.42	0.90	0.14	0.59
B3-E2	0.98	0.43	0.95	0.75	0.71
B3-E3	0.97	0.32	0.63	0.09	0.27

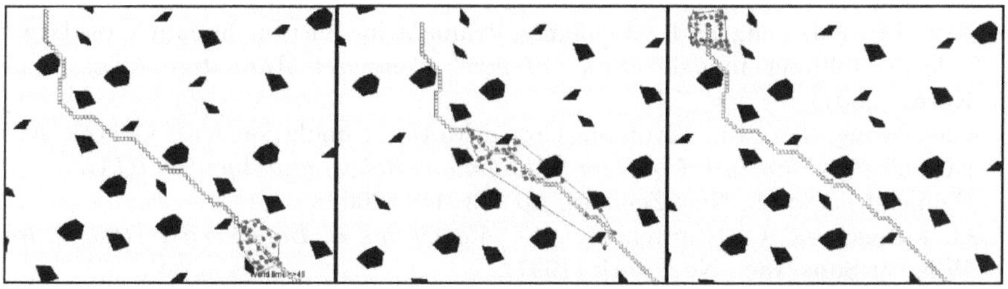

Fig. 14.1. Example of simulation results on the following behaviour (B2) with a small inter-robot distance in a space cluttered with small obstacles (E3)

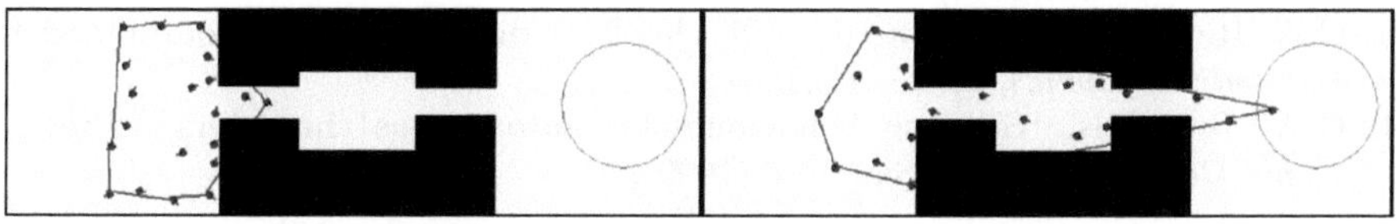

Fig. 14.2. Example of simulation results for the group moving behaviour (B1) with a large inter-robot distance passing a narrow passage (E2)

14.5 Conclusions

In this chapter, we have proposed to formulate the problem of how to compose virtual forces to drive the simulation as an optimization problem on the weighing parameters for virtual forces by the use of a genetic algorithm. The fitness functions used in the genetic algorithm are composed according to the desired behaviours from the elementary ones designed for evaluating a specific aspect of crowd motion. Our preliminary experiments reveal that the genetic algorithm is a good way to automate the time-consuming process of generating the optimal set of parameters for a given scene and a desired behaviour of movement.

References

1. T. Balch and R.C. Arkin, "Behaviour-based formation control for multirobot teams," *IEEE Trans. on Robotics and Automation*, 14(6), pp 926–939 (1998).
2. E. Bouvier, E. Cohen and L. Najman, "From crowd simulation to airbag deployment: particle systems, a new paradigm of simulation," *J. of Electronic Imaging*, 6(1), pp 94–107 (1997).
3. D.C. Brogan and J. Hodgins, "Group behaviors for systems with significant dynamics," *Autonomous Robots*, 4, pp 137–153 (1997).
4. J. Funge, X. Tu and D. Terzopoulos, "Cognitive model: knowledge, reasoning, and planning for intelligent characters," *Proc. of ACM SIGGRAPH*, pp 29–38, Los Angels (1999).

5. T.Y. Li, Y.J. Jeng and S.I. Chang, "Simulating virtual human crowds with a leader-follower model," *Proc. of 2001 Computer Animation Conf.*, Seoul, Korea (2001).
6. C.C. Wang, T.Y. Li. "Evolving Crowd Motion Simulation with Genetic Algorithm," *Proc. 3rd Intl. Conf. on Autonomous Robots and Agents - ICARA'2006*, Palmerston North, New Zealand, pp 443–448 (2006).
7. G. Mitsuo and C. Runwei, *Genetic Algorithms & Engineering Design*, John Wiley & Sons. Inc., New York (1997).
8. G.A. Miller. "The magical number seven, plus or minus two: Some limits on our capacity for processing information," *Psychological Review*, 63, pp 81–97 (1956).
9. S.R. Musse and D. Thalmann, "Hierarchical model for real time simulation of virtual human crowds," *IEEE Trans. on Visualization and Computer Graphics*, 7(2), pp 152–164 (2001).
10. C.W. Reynolds, "Flocks, herds, and schools: A distributed behavioural model," *Computer Graphics*, pp 25–34 (1987).
11. C.W. Reynolds, "Steering behaviours for autonomous characters," *Proc. of Game Developers Conf.*, San Jose (1999).

15

Application of a Particle Swarm Algorithm to the Capacitated Open Pit Mining Problem

Jacques A. Ferland[1], Jorge Amaya[2] and Melody Suzy Djuimo[1]

[1] Département d'informatique et de recherche opérationnelle
Université de Montréal, Montréal, Québec, Canada
ferland@iro.umontreal.ca, djuimoys@iro.umontreal.ca
[2] Departamento de Ingeniería Matemática and Centro de Modelamiento Matemático, Universidad de Chile, Santiago, Chile
jamaya@dim.uchile.cl

Summary. In the *capacitated open pit mining problem*, we consider the sequential extraction of blocks in order to maximize the total discounted profit under an extraction capacity during each period of the horizon. We propose a formulation closely related to the *resource-constrained project scheduling problem* (*RCPSP*) where the genotype representation of the solution is based on a priority value encoding. We use a *GRASP* procedure to generate an initial population (swarm) evolving according to a *particle swarm procedure* to search the feasible domain of the representations. Numerical results are introduced to analyze the impact of the different parameters of the procedures.

Keywords: Open pit mining, GRASP, Particle swarm, Evolutionary process, RCPSP, Priority encoding.

15.1 Introduction

Consider the problem where a mining industry is analyzing the profit of extracting the ore contained in some site. In fact, the analysis includes two stages where two different problems have to be solved. The first problem is to determine the *maximal open pit* corresponding to the maximal gain that the mining industry can get from the extraction at the site. This problem can be formulated as identifying the *maximal closure* of an associated oriented graph. Picard proposes a very efficient procedure to solve this problem in [14].

The second problem, denoted *block extraction problem*, is to determine the extraction order leading to the maximum profit accounting for the discount factor and particular constraints related to the extracting operation. In general, the mining site is partitioned into blocks characterized by several numbers. One of these is the net value of the block estimated from prospection

J.A. Ferland et al.: *Application of a Particle Swarm Algorithm to the Capacitated Open Pit Mining Problem*, Studies in Computational Intelligence (SCI) **76**, 127–133 (2007)
www.springerlink.com

data. This net value is equal to the profit associated with the block corresponding to the difference between the ore content value and the cost of extracting the block. Hence this value can be negative if the ore content value is smaller than the cost of extracting it. Furthermore, the physical nature of the problem may require extracting blocks having negative net values in order to have access to valuable blocks. In their pioneering work, Lerchs and Grossman [11] deals with this problem with an approach generating a sequence of nested pits. Then several other approach have been proposed: heuristics [1,8], Lagrangian relaxation [1], parametric methods [2, 6, 12, 17], dynamic programming [16], mixed integer programming [1,2,7], and metaheuristics or artificial intelligence methods [3, 4].

In this chapter, we rely on an analogy with the resource-constrained project scheduling problem (RCPSP) [9] to propose a *particle swarm procedure* [10] to solve a specific variant of the problem referred to as the *capacitated open pit mining problem*. Recall that a *particle swarm procedure* is an evolutionary population based process. The swarm (population) is evolving through the feasible domain searching for an optimal solution of the optimization problem. In the concluding remarks, we indicate how to extend it to other variants of this problem.

15.2 Solution Approach

The maximal open pit indicates the set of blocks to extract in order to maximize the total profit, accounting only for the constraints related to the *maximal pit slope* leading to identify a set of predecessors B_i including the blocks of the preceding layer above i that have to be removed before block i. But it does not give any indication of the order for extracting the blocs under operational constraints, nor does it account for the discount factor during the extraction horizon. This problem is more complex and hence more difficult to solve. In this chapter, we consider the capacitated open pit mining problem where only one additional operational constraint related to the maximal quantity that can be extracted during each period of the horizon. In our solution approach, we use the genotype representation $PR = \left[pr_1, \cdots, pr_{|N|}\right]$ that is similar to the priority value encoding introduced by Hartman in [13] to deal with the RCPSP. The ith element $pr_i \in [0,\ 1]$ corresponds to the priority of scheduling block i to be extracted. Hence the priority of extracting block i increases with the value of pr_i, and these values are such that $\sum_{i=1}^{|N|} pr_i = 1$.

Starting from the genotype vector PR, a feasible phenotype solution of the problem is generated using the following serial decoding scheme where the blocks are scheduled sequentially to be extracted. To initiate the first extraction period $t = 1$, we first remove the block among those having no predecessor (i.e., in the top layer) having the largest priority. During any period t, at any stage of the decoding scheme, the next block to be removed is one of those with the highest priority among those having all their predecessors already extracted such that the capacity C_t is not exceeded by its extraction. If no such block exists, then a new extraction period $(t + 1)$ is initiated. We denote by $v(PR)$ the value of the feasible solution associated with the genotype vector PR. To associate a priority to a block, we need to consider not only its net value b_i, but also its impact on the extraction of other blocks in future periods. One such measure proposed by Tolwinski and Underwood in [16] is the *block lookahead value* $\overline{b}_i$. This value is determined by referring to the *spanning cone* SC_i of block i including all blocks requiring that block i be removed before they can be extracted.

A particle swarm procedure [13] is evolving through the set of genotype vectors in order to search the feasible domain of the problem. Consider a population (swarm) P of M genotype vectors (particles): $P = \left\{PR^1, \cdots, PR^M\right\}$. The M initial genotype vectors are generated randomly using the GRASP to bias the values of the priorities in favour of the blocks having larger lookahead values. To initialize the first iteration of the particle swarm procedure, the genotype vector PR^k corresponds both to the current vector and the best achieved vector $\overline{PR}^k$ of the individual k (i.e., at the start of the first iteration, $\overline{PR}^k = PR^k$). During the iterations of the procedure, the individuals (genotype vectors) of the population are evolving, and we denote by PRb the best overall genotype vector achieved so far. Hence, at the end of each iteration, $\overline{PR}^k, k = 1, \cdots, M$, and PRb are updated as follows:

$$\overline{PR}^k := \text{ArgMax}\left\{v(PR^k),\ v(\overline{PR}^k)\right\}$$

$$PRb := \underset{1 \le k \le M}{\text{ArgMax}}\left\{v(\overline{PR}^k)\right\}.$$

Now we describe a typical iteration of the particle swarm procedure. Each current vector PR^k evolves individually to a new current genotype vector according to a probabilistic process accounting for its current value PR^k, for its best achieved value $\overline{PR}^k$, and for the best overall achieved vector PRb. More specifically, with each component i of each vector k, we associate a *velocity* [13] factor vc_i^k evolving at each iteration of the procedure. Its value is initialized at 0 (i.e., $vc_i^k = 0$) when the procedure starts. At each iteration, it evolves as follows:

$$vc_i^k := wvc_i^k + c_1 r_{1i}^k(\overline{pr}_i^k - pr_i^k) + c_2 r_{2i}^k(prb_i - pr_i^k),$$

and we define $ppr_i^k := vc_i^k + pr_i^k$.

The probabilistic nature of the procedure follows from the fact that $r_{1i}^k \in [0,\ 1]$ and $r_{2i}^k \in [0,\ 1]$ are independent uniform random number selected at each iteration. The impact of the terms is scaled by so called *acceleration coefficients* c_1 and c_2. The *inertia weight* w was introduced in [15] to improve the convergence rate.

Translating the vector $PPR^k = \left[ppr_1^k, \cdots, ppr_{|N*|}^k\right]$ to restore the non negativity of its component and normalizing the resulting vector, we determine the new current genotype PR^k.

15.3 Numerical Results

We are using 20 different problems randomly generated over a two dimensions grid having 20 layers and being 60 blocks wide. For each problem, 10 clusters including the blocks having ore inside are randomly generated (note that these clusters can overlap). The value b_i of the blocks belonging to the clusters is selected randomly in the set $\{6, 8, 12, 16\}$. The rest of the blocks outside the clusters have no ore inside, and they have a negative value b_i equal to -4. The weight p_i of each block i is equal to 1, and the maximal weight $C_t = 3$ for each period.

We use the first 10 problems having smaller maximal open pit ranging from 64 to 230 blocks to analyze the impact of the different parameters. Each problem is solved using 12 different combinations of parameter values where $\beta\% \in \{10, 50, 100\}$ ($\beta\%$ being the parameter in the GRASP procedure to increase the bias of the priorities in favour of the blocks having larger lookahead values as its value decreases), $M \in \{25, 50\}$, $w \in \{0.1, 0.7\}$, and $c_1, c_2 \in \{0.7, 1.4\}$.

For each combination of parameter values l, each problem ρ is solved five times to determine:

$va_{l\rho}$: the average of the best values $v(PRb)$ achieved
$vb_{l\rho}$: the best values $v(PRb)$ achieved
$\%_{l\rho}$: the average % of improvement $\%_{l\rho} = \frac{va_{l\rho} - v(PRb) \text{ at first iter.}}{v(PRb) \text{ at first iter.}} \times 100$

Then for each set of parameter l, we compute the average values va_l, vb_l, $\%_l$, and it_l over the 10 problems.

Impact of the size M of the population. The numerical results indicate that the values of the solutions generated are better when the size of the population is larger. This makes sense since we generate a larger number of different solutions.

Impact of the values of the particle swarm parameters. The numerical results indicate no clear impact of modifying the values of the parameters w, c_1, and c_2. Note that the values $w = 0.7$, $c_1 = 1.4$, and $c_2 = 1.4$ were selected accordingly to the authors in [5] who showned that setting the values of the parameters close to $w = 0.7298$ and $c_1 = c_2 = 1.49618$ gives acceptable results.

Impact of the $\beta\%$ in the GRASP procedure. From the numerical results, we observe that the values of va_l and vb_l decrease while the value of $\%_l$ increases as the value of $\beta\%$ increases. Since the priority of blocks with larger lookahead values get larger when $\beta\%$ decreases, we can expect the individuals

Table 15.1. Scenarios for the initial population

Scenario	#10	#2	#1	va_{sc}
1	50	0	0	277.53
2	49	0	1	250.97
3	39	10	1	250.64
4	25	24	1	250.96
5	10	39	1	250.55
6	0	0	50	247.12

in the initial population to have better profit values inducing that we can reach better solutions, but that the percentage of improvement $\%_l$ is smaller.

To analyze the impact of increasing the number of individuals in the initial population that are generated with smaller values of $\beta\%$, we use the last 10 problems having larger maximal open pit ranging from 291 to 508 blocks. Furthermore, we set the parameter values to those generating the best results for the first 10 problems: $M = 50$, $w = 0.7$, $c_1 = c_1 = 1.4$. But for $\beta\%$ we consider the six scenarios summarized in Table 15.1 where

#10: the number of individual generated with $\beta\% = 10$
#2: the number of individual generated with $\beta\% = 2$
#1: the number of individual generated with $\beta\% = 1$.

Each problem ρ is solved five times to determine:

$va_{sc\rho}$: the average of the best values $v(PRb)$ achieved for problem ρ using scenario sc
va_{sc}: the average value over the 10 problems.

Referring to Table 15.1, it is worth noticing that the generation of the initial population becomes more greedy (the number of individuals in the initial population generated with smaller values of $\beta\%$ increases) as the scenario number increases. The results in Table 15.1 indicates that in general (except for scenario 4) the value of va_{sc} decreases as the scenario number increases. Furthermore, the increases is quite important when we move from scenario 1 to scenario 2 and from scenario 5 to scenario 6. Hence it seems appropriate to use smaller values of $\beta\%$ (like 10) to generate the initial population of individuals, but also to avoid being too greedy using very small values of $\beta\%$.

How the solutions obtained with an evolutionary process like particle swarm compare with the one obtained by decoding a greedy genotype priority vector generated with the GRASP procedure where $\beta\% = 1$? The analysis is completed using the last 10 problems with the parameter values $M = 50$, $w = 0.7$, $c_1 = c_1 = 1.4$, and $\beta\% = 10$. Each problem ρ is solved five times to determine

$va_{1\rho}$: the average of the best values $v(PRb)$ achieved
$vb_{1\rho}$: the best value $v(PRb)$ achieved

Table 15.2. Improved solutions with particle swarm

Problem	$va_{1\rho}$	$vb_{1\rho}$	$vw_{1\rho}$	v_{greedy}
11	164.85	168.81	161.50	155.88
12	548.81	559.92	543.14	532.84
13	151.80	166.68	142.71	114.94
14	259.12	273.79	247.04	170.93
15	71.17	79.97	61.13	42.48
16	284.01	291.47	267.79	243.44
17	397.83	403.71	390.13	355.79
18	448.40	454.40	442.20	427.59
19	373.88	381.50	366.51	342.47
20	75.40	89.81	61.03	32.75

$vw_{1\rho}$: the worst value $v(PRb)$ achieved for parameter values in set 1
v_{greedy}: the value obtained by decoding a greedy genotype priority vector.

The results in Table 15.2 indicate that even the worst values $vw_{1\rho}$ is better than v_{greedy} for all problems. Furthermore, the percentage of improvement of $va_{1\rho}$ over v_{greedy} ranges from 1.03% to 56.56%. We can conclude that using particle swarm induces a gain in the solution quality.

Referring to this table, we can also observe that for each problem, the interval $[vw_{1\rho}, vb_{1\rho}]$ including the values $v(PRb)$ is quite small. Indeed, the smallest ratio $\Delta_\rho = \frac{vw_{1\rho}}{vb_{1\rho}}$ is equal to 0.809 for problem $\rho = 20$. This indicates that the particle swarm procedure is stable.

15.4 Conclusion

In this chapter, we consider the capacitated open pit mining problem to determine the sequential extraction of blocks maximizing the total discounted profit under an extraction capacity during each period of the horizon. We use a genotype representation of the solution based on a priority value encoding. An initial population (swarm) generated with a GRASP procedure is evolving according to a particle swarm procedure. The numerical results indicate that the better are the individuals in the initial population, the better is the solution generated. Also, the quality of the solution increases with the size of the population. The numerical results do not seem to be very sensitive to the parameter values of the particle swarm procedure. Finally, the quality of the solutions generated with the particle swarm procedure is better than those obtained with greedy genotype priority vectors.

References

1. Cacetta, L., Kelsey, P., Giannini, L.M. (1998), "Open pit mine production scheduling", *Computer Applications in the Mineral Industries International Symposium* (3rd *Regional APCOM*), *Austral. Inst. Min. Metall. Publication Series*, 5: 65–72.
2. Dagdelen, K., Johnson, T.B. (1986), "Optimum open pit mine production scheduling by lagrangian parameterization", *Proceedings of the 19*th *APCOM Symposium of the Society of Mining Engineers* (*AIME*), 127–142.
3. Denby, B., Schofield, D. (1995), "The use of genetic algorithms in underground mine scheduling", *Proceedings 25*th *APCOM Symposium of the Society of Mining Engineers* (*AIME*), 389–394.
4. Denby, B., Schofield, D., Bradford, S. (1991), "Neural network applications in mining engineering", *Department of Mineral Resources Engineering Mgzine*, University of Nootingham, 13–23.
5. Eberhart, R.C., Shi, Y. (2000), "Comparing inertia weights and construction factors in particle swarm optimization" *Proceedings of the Congress on Evolutionary Computation*, 84–88.
6. François-Bongarçon, D.M., Guibal, D. (1984), "Parametization of optimal designs of an open pit beginning a new phase of research", *Transactions SME, AIME*, 274: 1801–1805.
7. Gershon, M. (1983), "Mine scheduling optimization with mixed integer programming", *Mining Engineering*, 35: 351–354.
8. Gershon, M. (1987), "Heuristic approaches for mine planning and production scheduling", *International Journal of Mining and Geological Engineering*, 5: 1–13.
9. Hartmann, S. (1998), "A competitive genetic algorithm for the resource-constrained project scheduling", *Naval Research Logistics*, 456: 733–750.
10. Kennedy, J., Eberhart, R.C. (1995), "Particle swarm optimization", *Proceedings of the IEEE International Conference on Neural Networks, IV*: 1942–1948.
11. Lerchs, H., Grossman, I.F. (1965), "Optimum design for open pit mines", *CIM Bulletin*, 58: 47–54.
12. Matheron, G. (1975), "Le paramétrage des contours optimaux", *Technical report no. 403*, Centre de Géostatistiques, Fontainebleau, France.
13. Paquet, U., Engelbrecht, A.P. (2003), "A new particle swarm optimiser for linearly constrained optimization", *Proceedings of the 2003 Congress on Evolutionary Computation*, 227–233.
14. Picard, J.C. (1976), "Maximal closure of a graph and applications to combinatorial problems", *Management Science*, 22: 1268–1272.
15. Shi, Y., Eberhart, R.C. (1998), "A modified particle swarm optimizer", *Proceedings of the IEEE Congress on Evolutionary Computation, Piscataway, New Jersey*, 69–73.
16. Tolwinski, B., Underwood, R. (1996), "A scheduling algorithm for open pit mines", *IMA Journal of Mathematics Applied in Business & Industry*, 7: 247–270.
17. Whittle, J. (1998), "Four-X user manual", *Whittle Programming Pty Ltd*, Melbourne, Australia.

Automatic Adjustment for Optical Axes in Laser Systems Using Stochastic Binary Search Algorithm for Noisy Environments

Nobuharu Murata[1], Hirokazu Nosato[2], Tatsumi Furuya[1], and Masahiro Murakawa[2]

[1] Faculty of Science, Toho University, 2-2-1 Miyama, Funabashi, Chiba, Japan
nobu-murata@aist.go.jp, furuya@is.sci.toho-u.ac.jp
[2] ASRC, National Institute of Advanced Industrial Science and Technology (AIST), 1-1-1 Umezono, Tsukuba, Ibaraki, Japan
h.nosato@aist.go.jp, m.murakawa@aist.go.jp

Summary. We have proposed an automatic adjustment method using genetic algorithms (GA) to adjust the optical axes in laser systems. However, there are still two tasks that need to be solved: (1) long adjustment time and (2) adjustment precision due to observation noise. In order to solve these tasks, we propose a robust and efficient automatic adjustment method for the optical axes of laser systems using stochastic binary search algorithm. Adjustment experiments for optical axes with 4-DOF demonstrate that the adjustment time could be reduced to half of conventional adjustment time with GA. Adjustment precision was enhanced by 60%.

Keywords: automatic adjustment, optical axes, genetic algorithm, binary search, noisy environments.

16.1 Introduction

Laser systems are currently essential in various industrial fields. For laser systems, the adjustment for the optical axes is crucial, because the performance of the laser system deteriorates when the optical axes deviate from their specification settings, due to disturbances such as vibrations. However, it is very difficult to adjust the optical axes, because adjustment requires high-precision positioning settings with micrometer resolutions and because it is necessary to adjust for multi-degree-of-freedom (DOF) that have an interdependent relationship. Thus, adjustment costs are a major problem due to the huge amount of time required for a skilled engineer to adjust the optical axes. In order to overcome this problem, we have proposed automatic adjustment methods for optical axes using genetic algorithms (GA) [1–3]. For example, our method has been successfully applied to the automatic adjustment of a femto-second laser

N. Murata et al.: *Automatic Adjustment for Optical Axes in Laser Systems Using Stochastic Binary Search Algorithm for Noisy Environments*, Studies in Computational Intelligence (SCI) **76**, 135–144 (2007)
www.springerlink.com

that has 12-DOF [2]. However, there were two problems with the proposed methods that needed to be solved. First, it has been necessary to reduce the adjustment time to within 10 min. Because a laser system should ideally be readjusted every time it is used. For practical considerations, adjustment time must be as fast as possible. Second, because the adjustment of the optical axes is usually undertaken in very noisy environments, the precision of adjustment can vary widely.

In order to overcome these problems, we propose a novel adjustment method which has two characteristics (1) The method adopts a Binary Search Algorithm (BSA) [4]. The BSA gradually changes from the exploration phase to the exploitation phase. The exploration phase searches a region that has not previously been searched using a binary search tree. The exploitation phase searches a region around good points. (2) The fitness value adopts weighted average of sampled fitness values using a search history.

There are two advantages with the proposed method: (1)adjustment time can be reduced. The method does not search in regions that have previously been searched. In addition, it is not necessary to reevaluate the fitness function to mitigate the influence of noise. (2)It provides robust automatic adjustment. Instance of premature convergence or falling into local solutions do not occur because the adjustment is less influenced by noise. Accordingly, we can realize robust and efficient automatic adjustment systems for the optical axes within laser system by the proposed method. Conducted experiments involving 4-DOF adjustment with the proposed method demonstrate that (1)adjustment time can be reduced to half of conventional adjustment time, and (2)precision can be enhanced by 60%.

This chapter is organized as follows: In Sect. 16.2, we explain the adjustment system for the optical axes of laser systems and the conventional method of automatic adjustment. Section 16.3 describes our proposed method, and Sect. 16.4 outlines the automatic adjustment system used in the experiments. In Sect. 16.5, we present the experimental results obtained for the proposed method. Finally, a summary of this study and future investigations are provided in Sect. 16.6.

16.2 Adjustment Systems for Optical Axes and Automatic Adjustment Methods

16.2.1 Adjustment Systems for Optical Axes

For laser systems, the adjustment of the optical axes is crucial because the performance of laser systems deteriorates when the optical axes deviate from their specification settings. Let us explain adjustment systems taking a femto-second laser system as an example. A characteristic of femto-second lasers is that the high-peak power is inversely proportional to the short duration of the laser pulses, so they can generate high power levels, over 1 MW, during femtosecond (10^{-15} s) pulses. This system that consists of seven mirrors and two prisms has 12-DOF to be adjusted with micrometer resolutions. The optical axes are adjusted by moving stepping motors, so that the output power from the laser system is maximized.

In such an adjustment system, there are two sources of observational noise influencing output evaluation. The first is the noise in the detectors that evaluates the output from the laser system.The second source is the precision of the positioning motors. While the positioning motors are moved according to constant displacements, actual axial displacements are not constant. Therefore, the optical axes can deviate from the desired state, even if motors are moved according to the displacement settings in seeking to adjust the target state. Thus, the adjustment of optical axes must be carried out by considering these sources of noise. In the system, for example, manual adjustment takes about a week.

16.2.2 Automatic Adjustment Methods

We have already demonstrated how it is difficult for a hill climbing method to automatically adjust optical axes [2]. There are two reasons for this. The first reason is that the adjustment becomes trapped by local solutions, because the adjustment points of the laser system have an interdependent relationship. The second reason is that adjustment must be executed in noisy environments. In order to overcome these problems, we have proposed an adjustment method using GA [1–3]. In the proposed method, a chromosome is a set of genes, which represent motor displacements, and fitness is the output from the laser system. We have demonstrated the effectiveness of automatic adjustment using GA for the laser system.However, there were two problems that needed to be solved: (1) Adjustment took a long time, because the method also performed exploration during the final phase. The time for motor movements, which accounts for nearly all of the adjustment time, increases in proportion to the degree of motor displacement. (2) Robust adjustment is difficult, because search is influenced by the two sources of noise explained in sect. 16.2.1. Consequently, instances of premature convergence occurred or adjustment became trapped to local solutions, so adjustment precision varied widely.

16.3 Proposed Method

We propose a robust and efficient automatic adjustment for the optical axes of laser system using a binary search algorithm (BSA) [4] for noisy environments. The flowchart of the proposed method is shown in Fig. 16.1. This method utilizes weighted averaged fitness [5] in the BSA as explained in Fig. 16.1. We refer to the proposed method as BSW. The algorithm is explained in more detail later.

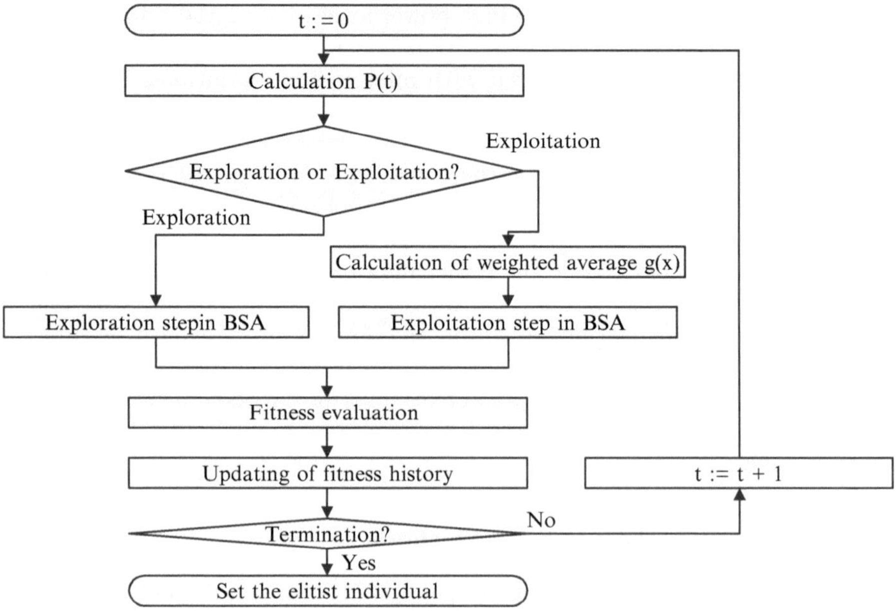

Fig. 16.1. Flowchart for the proposed method

16.3.1 Binary Search Algorithm

The strategy of BSA is to use a binary search tree [4] to divide the search space into empty regions, allowing for the largest empty regions to be approximated. The search tree is constructed by generating a point x_t at random within the chosen hypercube, then by dividing the hypercube along the dimension that yields the most "cube-like" subspaces. The basic algorithm for constructing the binary search tree works by repeatedly choosing an exploration or exploitation step:

1. Exploration: Next point x_{t+1} is generated within the largest empty region.
2. Exploitation: Next point x_{t+1} is generated within the largest empty region that is within a small distance from a "good point."

The coordinates of the point x_t and the evaluated value f_t at x_t are stored in a search history $F(t)$ represented in (16.1).

$$F(T) = \{(x_1, f_1), (x_2, f_2), \ldots, (x_i, f_i), \ldots, (x_T, f_T)\} \tag{16.1}$$

The decision of whether to perform exploration or exploitation is made based on a probability distribution $P(t)$ that varies with the number of fitness evaluations. $P(t)$ calculated using (16.2) is illustrated graphically in Fig. 16.2.

$$P(t) = (C - 1)\frac{\tanh(\frac{t/N-K}{\sigma}) - \tanh(\frac{-K}{\sigma})}{\tanh(\frac{1-K}{\sigma}) - \tanh(\frac{-K}{\sigma})} + 1 \tag{16.2}$$

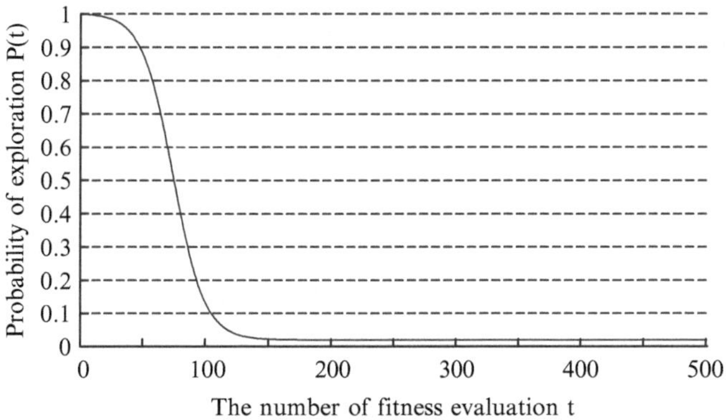

Fig. 16.2. Probability of exploration for C=0.02, K=0.1, σ=0.05, and N=500

C is the minimum probability of performing the exploration step, σ is the rate at which the probability of exploration decays, K is the mid point of the decay, and N is the maximum number of trials that are to be performed.

The automatic adjustment method using BSA can efficiently adjust optical axes, because the search phase in BSA gradually shifts from exploration to exploitation according to $P(t)$ based on the number t of fitness evaluation, as shown in Fig. 16.2.

16.3.2 Weighted Averaged Fitness

The conventional method of coping with noisy fitness functions is to evaluate the fitness values several times for each individual and to adopt the average of the sampled values [6, 7]. However, adjustment for laser systems by the conventional method is not practical, because adjustment time increases. For example, if moving the motors and detecting to obtain a fitness value are performed N times, then the adjustment time increases N-fold.

In order to solve this problem, a weighted average value, which is calculated from the search history, is used for the fitness value. For the laser system, we assume the detected value f_t at x_t increases or decreases in proportion to a distance d_t from a certain point x to the point x_t and that noise varies according to a normal distribution. The maximum likelihood estimation of $f(x)$ can be obtained as follows:

$$g(x) = \frac{f(x) + \sum_{t=2}^{T} \frac{1}{1+k \times d_t^2} f_t}{1 + \sum_{t=2}^{T} \frac{1}{1+k \times d_t^2}}, \tag{16.3}$$

$$d_t = \|x - x_t\|, \tag{16.4}$$

where $f(x)$ is evaluated value with the detector, $f_1 = f(x)$, $x_1 = x, k$ is the proportional value, and d_t is the distance from sampled points. The "good point" in exploitation step of the BSA is decided using this weighted averaged fitness $g(x)$.

There are two advantages of this method. The first is that this method can prevent premature convergence during the exploitation phase due to observational noise. The second advantage is that the number of evaluation is just one time for each individual. Thus, this method is capable of adjusting the optical axes robustly and efficiently in noisy environments.

16.4 An Automatic Adjustment System for Optical Axes

In this chapter, we demonstrate the effectiveness of the proposed method using the most basic adjustment system [3, 5]. This adjustment system consists of two mirrors with two stepping motors, an evaluation detector to detect the positioning of the optical axis, a motor controller to control the stepping motor for the mirror, and PC to execute the flow explained in Fig. 16.1. The mirrors in the system can be adjusted according to 2-DOF to adjust the positioning of the optical axis.

A photograph of the experimental system is shown in Fig. 16.3. The motor controller consists of the stepping motor driver and the mirror holder system with a resolution of 0.075 μm/step. With this controller, the time to move the motors in evaluating each individual is at maximum of about 4 s. The detector with a resolution of 1.39 μm, can detect X–Y coordinates of the optical axes on the detector front. The pumping source is a *He–Ne* gas laser and the beam diameter is 2 mm.

The optical axes are adjusted automatically based on the flow shown in Fig. 16.1. A chromosome in the BSW represents displacement for the four stepping motors, which moves the two mirrors. The displacement of each

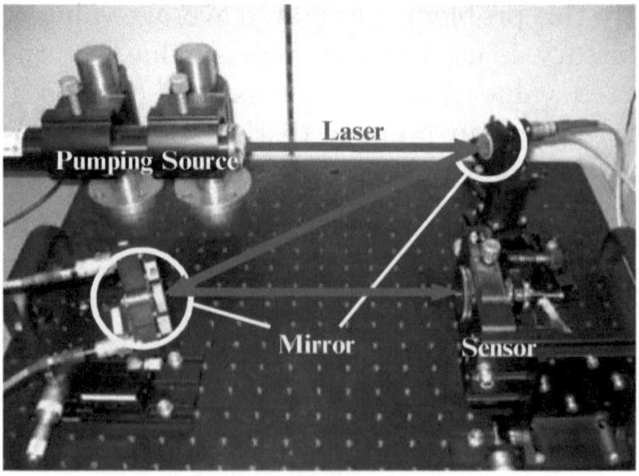

Fig. 16.3. The automatic adjustment system for optical axes with 4-DOF

stepping motor is represented by 8 bits. The evaluation value $f(x)$ uses a positioning error, which is the Euclidean distance between the positioning of the optical axes and the target position, calculated from the detected X–Y coordinates.

16.5 Adjustment Experiments

16.5.1 The Details of Experiments

Three experiments were conducted to examine the effectiveness of the proposed method. These were an automatic adjustment experiment using a GA, automatic adjustment experiment using BSA, and the automatic adjustment experiment using BSW.

The adjustment goal in each case was set to the laser system to its ideal state (i.e., the error in terms of the target positioning of the optical axes is 0). The initial conditions were random state where the positioning was altered within a range of ±5 mm. Adjustment started from the initial states. Adjustment terminated when the number of fitness evaluations was 500, and were conducted over 10 trials. After adjustment, the optical axes were set to the best state based on the identified elitist individual. The parameters in these experiments were as following:

- GA: The population size was 20 and the probabilities for crossover and mutation were 0.7 and 0.05, respectively.
- BSA: Parameters of $P(t)$ were $C = 0.02$, $K = 0.15$, $\sigma = 0.05$, and $N = 500$.
- BSW: Parameters of $P(t)$ were the same as BSA, and $k = 100$.

16.5.2 Experimental Results

Table 16.1 presents the average results for 10 trials for each adjustment. Each trial result is shown in Table 16.2. In these table, "fit" refers to fitness and "reset" is the evaluation value, i.e., positioning error, when the optical axis is reset to the best state based on the adjusted results. In these tables, the fitness values for the GA and BSA are actual detected values, while the fitness value of BSW is a calculated weighted average value, which is indicated by * in Tables 16.1 and 16.2.

Table 16.1. The results for three adjustments

	Time	Fit-ave	Fit-σ	Reset-ave	Reset-σ	Positioning error-ave (μm)
GA	26.1	17.7	11.2	31.3	16.8	43.5
BSA	10.6	10.7	16.0	28.9	19.8	40.2
BSW	13.1	22.7*	12.1*	13.5	10.4	18.8

Table 16.2. The results for each trial of the averaged results for the three adjustments

	GA		BSA		BSW	
	Fitness	reset	Fitness	reset	Fitness	reset
1st trial	14.0	7.6	1.4	10.6	14.0*	4.5
2nd trial	20.0	19.8	7.0	19.0	16.5*	13.0
3rd trial	9.4	40.3	53.6	54.9	13.5*	16.3
4th trial	28.3	51.6	8.5	63.0	11.2*	4.1
5th trial	17.2	22.1	1.0	17.1	46.7*	38.3
6th trial	28.0	24.7	11.4	9.1	23.1*	7.3
7th trial	38.1	49.9	18.0	30.5	21.5*	13.9
8th trial	9.2	56.2	4.1	43.0	26.6*	7.3
9th trial	3.2	23.0	1.0	7.1	40.7*	22.5
10th trial	9.0	17.7	1.4	35.2	13.8*	8.1
Average	17.7	31.3	10.7	28.9	22.7*	13.5
Standard deviation(σ)	11.2	16.8	16.0	19.8	12.1*	10.4

First, looking at the results for the GA and BSA, clearly both could adjust the optical axes, because the fit-average values were 17.7 ($17.7 \times 1.39 = 24.6\,\mu\text{m}$) and 10.7 ($10.7 \times 1.39 = 14.9\,\mu\text{m}$) for the GA and BSA, respectively. Moreover, efficient adjustment was achieved by BSA, because, in terms of adjustment times, the BSA required 10.6 min while the GA took 26.1 min. The reason for this difference is that BSA does not carry out the exploration step in the final phase. However, the levels of precision with GA and BSA deteriorated after resetting of the optical axes: the precision for the GA dropped to 31.3 ($31.3 \times 1.39 = 43.5\,\mu\text{m}$) from 17.7 ($24.6\,\mu\text{m}$), while the precision for the BSA dropped to 28.9 ($28.9 \times 1.39 = 40.2\,\mu\text{m}$) from 10.7 ($14.9\,\mu\text{m}$). The reason for this is that both the GA and BSA are influenced by observation noise. The large reset-σ values are consistent with this explanation, as shown in Table 16.1. Therefore, both BSA and GA failed to provide robust adjustment.

Second, turning to compare the results for the BSA and BSW, in terms of the average value after the optical axes were reset, the precision with the BSW improves quite remarkably in contrast with that for the BSA, because BSA was 28.9 ($40.2\,\mu\text{m}$) while BSW was 13.5 ($13.5 \times 1.39 = 18.8\,\mu\text{m}$), as shown in Table 16.1. This results indicates that the precision was enhanced by 60% compared to the results for the GA. Moreover, the BSW was able to achieve a robust adjustment, because reset-σ value for the BSW was 10.4 while reset-σ value for the BSA was 19.8. The BSW was less influenced by noise than the BSA, because it adopted a weighted average value for the fitness value. In terms of adjustment times, the BSW could adjust the optical axes within half the time required for the GA, because the BSW required only 13.1 min while GA took 26.1 min. However, adjustment using BSW took a little more time compared to the BSA, because of the time involved in calculating the weighted average values, although this can be improved by limiting the extent of search history used.

16.6 Conclusions

In this chapter, we have proposed a method for the robust and efficient automatic adjustment of optical axes using BSW. From the results of adjustments experiments using the proposed method, we have demonstrated that adjustment could be performed with half the time and with enhanced precision compared to the conventional method by 60%. Our proposed method achieved robust and efficient automatic adjustment of the optical axes.

There are three aspects of this method that require future investigation. The first aspect is the application of the proposed method to optical systems involving multiple components with several D.O.F to verify its effectiveness for more difficult adjustment tasks. The second aspect is the extension of the proposed method in order to adjust the optical axes of laser systems with multiobjective [3], because the adjustment of optical axes needs essentially multiobjective adjustment. The third aspect is to reduce the levels of calculation in the weighted average values, because most of the overhead cost in our method is the additional time to calculate the weighted average values. We are aiming to demonstrate the effectiveness of the proposed method in high-end laser systems, such as femto-second lasers.

Acknowledgments

This work was supported in 2004 by Industrial Technology Research Grant Program of the New Energy and Industrial Technology Development Organization (NEDO) of Japan and Grant-in-Aid for JSPS Fellows in 2005.

References

1. M. Murakawa, T. Itatani, Y. Kasai, H. Yoshikawa, and T. Higuchi. An evolvable laser system for generating femtosecond pulses. *Proceedings of the Second Genetic and Evolutionary Computation Conference (GECCO 2000)*, Las Vegas, pages 636–642, (2000).
2. H. Nosato, Y. Kasai, M. Murakawa, T. Itatani, and T. Higuchi. Automatic adjustments of a femtosecond-pulses laser using genetic algorithms. *Proceedings of 2003 Congress on Evolutionary Computation (CEC 2003)*, Canberra, Australia, pages 2096–2101, (2003).

3. N. Murata, H. Nosato, T. Furuya, and M. Murakawa. An automatic multi-objective adjustment system for optical axes using genetic algorithms. *Proceedings of the 5th International Conference on Intelligent Systems Design and Applications (ISDA 2005)*, Wroclaw, Poland, pages 546–551, (2005).
4. E.J. Hughes. Multi-objective binary search optimisation. *Proceedings of the Second International Conference on Evolutionary Multi-Criterion Optimisation (EMO 2003)*, Faro, Portugal, pages 102–117, (2003).
5. N. Murata, H. Nosato, T. Furuya, and M. Murakawa. Robust and Efficient Automatic Adjustment for Optical Axes in Laser Systems using Stochastic Binary Search Algorithm for Noisy Environments *Proceedings of the 3rd International Conference on Autonomous Robots and Agents (ICARA 2006)*, Palmerston North, New Zealand, pages 261–266, (2006).
6. J.M. Fitzpatrick and J.J. Greffenstette. Genetic algorithms in noisy environments. *Machine Learning*, 3:101–120, (1988).
7. P. Stagge. Averaging efficiently in the presence of noise. *Proceedings of Parallel Problem Solving from Nature (PPSN V)*, Amsterdam, pages 188–197, (1998).

17

An Analysis of the Chromosome Generated by a Genetic Algorithm Used to Create a Controller for a Mobile Inverted Pendulum

Mark Beckerleg and John Collins

Engineering Research Institute
Auckland University of Technology
Auckland, New Zealand
m.beckerleg@aut.ac.nz, jcollins@aut.ac.nz

Summary. This paper presents a novel method which uses a genetic algorithm to evolve the control system of a computer simulated mobile inverted pendulum. During the evolutionary process the best and average fitness, the chromosome and pendulum motion is recorded and analysed. The chromosome is a two dimensional lookup table that relates the current pendulum's angle and angular velocity to the motor direction and torque settings required to maintain balance of the pendulum. The lookup table is modified by a genetic algorithm using tournament selection, two-point crossover and a mutation rate of 2%. After two hundred generations the evolved chromosome is capable of balancing the pendulum for more than 200 s.

Keywords: Evolutionary computation, genetic algorithm, mobile inverted pendulum.

17.1 Introduction

Mobile inverted pendulums have moved from university research to commercial reality with the advent of the Segway. The reduction in costs of gyroscopes and accelerometers has enabled universities and students to use these in project applications. The use of these devices and the pendulum model can also be modified for other fields such as bipedal locomotion [1] or game playing such as soccer [2].

Mathematical models describing the behaviour of the pendulum have been developed enabling the construction of PID controllers which provide excellent balancing performance [3, 4]. Other researchers have investigated fuzzy logic [5–7] and neural network [8, 9] controllers to maintain the pendulum balance. Genetic algorithms have not previously been used to control a mobile inverted pendulum.

M. Beckerleg and J. Collins: *An Analysis of the Chromosome Generated by a Genetic Algorithm Used to Create a Controller for a Mobile Inverted Pendulum*, Studies in Computational Intelligence (SCI) **76**, 145–152 (2007)
www.springerlink.com

This paper is an extension of a paper first published in ICARA 2006 [10] which investigates the techniques and associated problems of evolutionary computation and genetic algorithms to evolve the controller for the pendulum.

The main difficulties in the field of evolutionary robotics are: (1) initial chromosomes are destructive to the robot and its environment; (2) initial chromosome populations have very little selective pressure (bootstrap problem [11]); (3) robotic tasks are complex creating a large search space and time required to evolve a controller. Robotic simulation is used to overcome the first two problems. The latter can be diminished by either reducing the chromosome size and thus the search space [12], or by using subsumption architecture where individual behaviours of the robot are evolved independently of each other before being combined together [13, 14]. In this application we use subsumption architecture in which the first behaviour to evolve is balancing. Future behaviours such as navigation or autonomy can then be independently evolved.

The Auckland University of Technology has constructed a mobile inverted pendulum for this research project which is capable of balancing using standard PID control software. The hardware is capable of supporting either an Atmel 8 bit microcontroller, an ARM 32 bit processor or an Altera Cyclone II FPGA allowing research into evolutionary computation, evolvable artificial neural networks, evolvable fuzzy logic and evolvable hardware. The first stage of the research is to create a simulation of the pendulum to allow testing of genetic algorithms.

17.2 Mathematical Modelling

The pendulum is a non-holonomic robot with three degrees of freedom (DOF), two planar motions and one tilt-angular motion, but with direct control in only the planar motions. Thus the control of the planar motion must work in such a way as to control the tilt/angular motion. As shown in Fig. 17.1, the

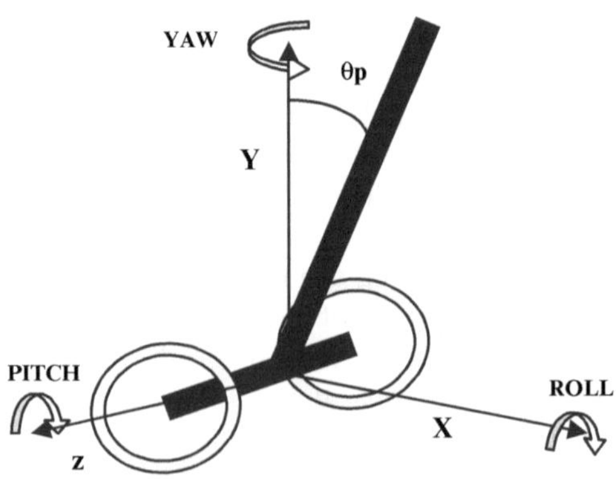

Fig. 17.1. Physical model of a pendulum

pendulum can rotate around the z axis (tilt), this is described by its angle θp and its angular velocity ωp. The pendulum can move on its x axis described by its position x and its velocity v. The pendulum can also rotate around the y axis described by the angle Φ and angular velocity Φ'.

To simplify the evolutionary process, the pendulum will be constrained to only one planar axis x by driving the two wheels together so there will be no yaw, and the pendulum is on a flat surface so there will be no roll.

The important parameters are the pendulum's angle, angular velocity and the linear displacement along the x-axis. It can be shown that the motion of the pendulum can be described by equations (17.1) and (17.2).

$$\left(M_p + 2M_w + \frac{2I_w}{r^2}\right)\ddot{x} + M_p l \cos\theta_p \ddot{\theta}_p = \frac{2T}{r} + M_p l \dot{\theta}_p^2 \sin\theta_p \quad (17.1)$$

$$M_p l \cos\theta_p \ddot{x} + I_p \ddot{\theta}_p = M_p g l \sin\theta_p - 2T \quad (17.2)$$

The computer simulation solves these equations in quantized time intervals to determine the pendulum's current horizontal velocity, horizontal displacement, angle and angular velocity.

17.3 Genetic Algorithm

A chromosome is a possible solution to a problem. In regard to the pendulum, the chromosome is the required motor direction and torque for a given angle and angular velocity. The motor driver for the pendulum is an H-bridge driver controlled by an eight bit number. The number is a linear representation of the motor direction and torque, with 0 representing the maximum reverse torque, 125 is the motor stopped and 250 is the maximum forward torque. The step size is 25 allowing eleven possible torque settings.

The chromosome is a two dimensional lookup table which relates the angle and angular velocity of the pendulum to the required motor direction and torque. The rows represent the pendulum's angle ranging $\pm 18°$ from vertical with a step size of $3°$. The columns represent the pendulum's angular velocity ranging ± 30 degrees per second with a step size of 5 degrees per second. The population size is 100 with the initial population randomly generated.

Chromosome reproduction creates two offspring per pair of parents, using a two point crossover scheme where two points within the chromosome are randomly chosen and the gene code within these two points is swapped between the two parents. To maintain population diversity two percent of the population is mutated, where each mutated chromosome has ten genes randomly altered.

The selection process used is tournament selection in which a subgroup of individuals is selected from the population and only the fittest individual within the subgroup is retained. The selection pressure is increased as the subgroup size increases because fewer individuals are being kept in each generation, however the gene pool will rapidly decrease with a high selection pressure, increasing the possibility of the evolutionary process becoming stuck in local maxima. For this experiment the subgroup size is two, which has poor selection intensity but maintains the widest gene pool.

The fitness is determined by the length of time that the pendulum remains within a vertical range of ±18° and remains within ±0.5 m of its start position of 18° with an angular velocity of zero. Each fitness test is stopped after 300 s if the pendulum has not gone outside the maximum vertical range within this time.

17.4 Simulation

The simulation shown in Fig. 17.2 has a diagrammatic representation of the pendulum and numerical displays of the pendulum's parameters. The pendulum starting angle, current generation, individual, average fitness and maximum fitness are also displayed. These parameters are saved to a file for future analysis. The simulation can be set to run in real time allowing the motion of the pendulum to be observed, or it can be sped up allowing faster evolution but then the pendulum cannot be observed.

The best individual's fitness, chromosome and motion were recorded. The motion showed the positions in the array that the individual stepped through as it was tested for fitness. This data allows us to monitor how the chromosome is evolving, to see what parts of the array the individual passed through, where it spent most of its time and what caused the individual to fail the test (angle, horizontal displacement or timeout).

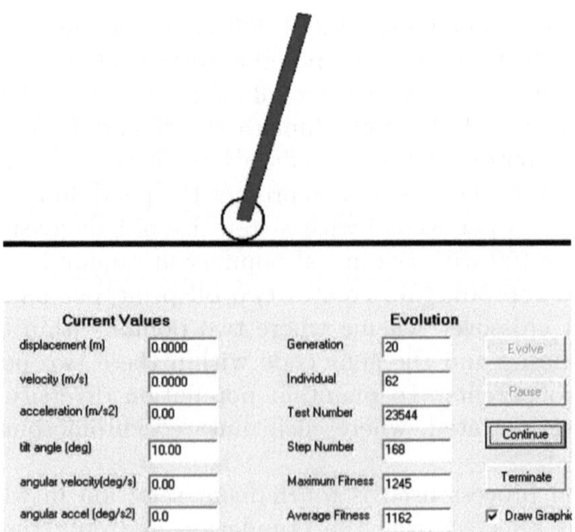

Fig. 17.2. Simulation GUI

When the simulation was first run it was found that under certain conditions the controller would stop evolving. To prevent this both the simulation and the fitness criteria needed to be modified.

It was found only a limited part of the controller array would evolve. This is because the pendulum would learn to balance from its initial offset starting angle. However the pendulum never evolved how to balance from other start conditions. To overcome this, the same individual was tested over a range of starting angles with a starting angular velocity of zero.

It was found that the pendulum would maintain a balance point with a constant horizontal motion. This characteristic was penalized by terminating the test when the pendulum moved $\pm 0.5\,\mathrm{m}$ from the initial horizontal starting position. Thus the fitness for each individual was determined by both its ability to balance the pendulum and to remain stationary.

It was found that some tests would last an indefinite period of time. This is a good result for one individual at one starting point, however the fitness evaluation would never end and the evolution process would stop. To overcome this all individuals are terminated after a 300-s interval.

It was found that the extreme positions of the array did not evolve to the value expected from the mathematical model, i.e. moving to a maximum forward or reverse torque. This is because at these positions the pendulum is not able to maintain balance, even with the motor at maximum torque. Thus parts of the array will not evolve as there are no values that will offer a significant difference in fitness.

It was found that the motor torque was not strong enough to pull the pendulum upright within the $\pm 0.5\,\mathrm{m}$ boundary for the start angles of ± 18 and ± 15, thus the best possible fitness is about 220 s.

17.5 Results

A typical result is shown in Fig. 17.3 with an initial large gap between the best and average fitness, and then a convergence between the average and best fitness with each generation. The best fitness increases in steps where as the average fitness gradually improves. An evolved chromosome for a recorded run is shown in Table 17.1. Each individual was tested over the range of start angles and each step that the pendulum took in the lookup table was recorded.

Initially the simulation was tested with a chromosome from a mathematical model of the pendulum. This model had a linear progression of motor torque dependant on the angle and angular velocity. When this chromosome was tested it performed poorly, failing the test after less than one second. A review of the pendulum's motion showed the reason for failure was that even though the pendulum moved to an upright position it had horizontal drift outside the $\pm 0.5\,\mathrm{m}$ distance. From this it can be seen that successful chromosomes require torque settings that quickly move the pendulum to a vertical position and then keep it within the horizontal boundary.

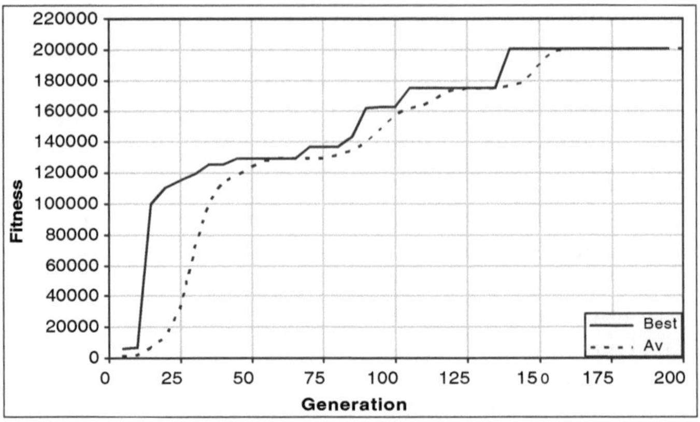

Fig. 17.3. Average and best fitness

Table 17.1. Evolved chromosomes

	Angular velocity												
Angle	**−30**	**−25**	**−20**	**−15**	**−10**	**−5**	**0**	**5**	**10**	**15**	**20**	**25**	**30**
−18	50	100	225	0	50	175	25	100	50	150	200	50	125
−15	25	250	25	125	0	0	50	125	75	175	250	225	100
−12	175	150	25	25	25	225	75	75	0	175	25	100	75
−9	250	25	25	25	150	0	100	75	0	200	50	50	100
−6	25	125	150	75	150	175	75	25	125	0	100	50	0
−3	100	175	0	150	25	50	0	25	25	0	100	25	225
0	225	100	100	150	25	250	75	150	200	50	100	200	100
3	225	75	75	125	100	250	225	100	225	100	250	0	100
6	200	225	200	175	225	225	0	200	175	100	0	150	25
9	250	150	200	100	175	0	225	125	200	25	175	150	100
12	225	225	175	200	225	125	225	225	150	200	25	225	150
15	250	250	250	225	125	250	200	225	125	125	225	250	250
18	50	250	50	50	75	150	250	250	100	0	175	200	200

An analysis of an evolved chromosome and the sequence that it moves through the chromosome shows the pendulum is quickly brought upright and then jitters around a position so that horizontal drift is eliminated. The pendulum eventually runs to the timeout limit or a small horizontal drift takes the pendulum to the positional limit. The evolved chromosome has a non-linear progression between cells which allows it to become vertical on start up and jitter around an angle of zero degrees.

The path that the pendulum takes through the chromosome is highly convoluted. Some of the cells would seem to have incorrect values according to their angle and angular velocity. However adjacent cells compensate for the incorrect settings. It is this erratic sequence of motor torques that creates the jitter and corresponding horizontal stability.

The best chromosomes from several evolutionary runs were compared and it was found that the chromosomes were quite different even though they had a similar fitness. This is because the initial chromosomes are random and the convoluted pathways that they evolve are unique to each starting individual.

17.6 Conclusions

This paper has described a method in which a simulation of a mobile inverted pendulum controller can be evolved to a point where an acceptable level of balance is achieved. It was found that the GA produced a controller capable of balancing the pendulum for 200 s within 200 generations. The next stage of research is to move the application from simulation to the AUT mobile inverted pendulum.

References

1. K.-K. Noh, J.-G. Kim, and U.-Y. Huh, "Stability experiment of a biped walking robot with inverted pendulum," Industrial Electronics Society, 2004. (IECON) 2004. 30th Annual Conference of IEEE, Busan, Korea, 3, pp 2475–2479 (2004).
2. B. Browning, P.E. Rybski, J. Searock, and M.M. Veloso, "Development of a soccer-playing dynamically-balancing mobile robot," Proceedings IEEE International Conference on Robotics and Automation, New Orleans, pp 1752–1757 (2004).
3. F. Grasser, A. D'Arrigo, S. Colombi, and A.C. Rufer, "JOE: a mobile, inverted pendulum," Industrial Electronics, IEEE Transactions on, vol. 49, pp 107–114, (2002).
4. Y. Kim, S. Kim, and Y. Kwak, "Dynamic Analysis of a Nonholonomic Two-Wheeled Inverted Pendulum Robot," Journal of Intelligent and Robotic Systems, vol. 44, pp 25–46, (2005).
5. X.M. Ding, P.R. Zhang, X.M. Yang, Y.M. Xu, "The application of hierarchical fuzzy control for two-wheel mobile inverted pendulum," Electric Machines and Control, vol. 9, pp 372–5, (2005).

6. S.S. Ge, Y.K. Loi, and C.-Y. Su, "Fuzzy logic control of a pole balancing vehicle," Proceedings of the 3rd International Conference on Industrial Automation, Montreal, pp 10–17, (1999).
7. L. Ojeda, M. Raju, and J. Borenstein, "FLEXnav: a fuzzy logic expert dead-reckoning system for the Segway RMP," Proceedings Unmanned Ground Vehicle Technology VI, International Society for Optical Engineering, Orlando, FL, USA, pp11–23 (2004).
8. S.S. Kim, T.I. Kim, K.S. Jang, and S. Jung, "Control experiment of a wheeled drive mobile pendulum using neural network," Proceedings 30th Annual Conference of IEEE Industrial Electronics Society (IECON 2004), 2–6 Nov. 2004, Busan, South Korea, pp 2234–2239 (2004).
9. T.-C. Chen, T.-J. Ren, and C.-H. Yu, "Motion control of a two-wheel power aid mobile," Proceedings SICE Annual Conference 2005, Aug 8–10 2005, Okayama, Japan, (2005).
10. M. Beckerleg and J. Collins, "A GA Based Controller for a Mobile Inverted Pendulum," 3rd International Conference on Autonomous Robots and Agents (ICARA 2006), pp 123–127, 12–14 December, Palmerston North, New Zealand, (2006).
11. A.L. Nelson and E. Grant, "Developmental Analysis in Evolutionary Robotics," IEE mountain workshop on Adaptive and Learning Systems, pp 201–206 (2006).
12. I. Harvey, "Artificial Evolution: A Continuing SAGA," Proceedings Evolutionary Robotics : from intelligent robotics to artificial life (ER 2001), International symposium, Tokyo, JAPAN, 2217, pp 94–109 (2001).
13. R.A. Brooks, "A robot that walks; emergent behaviours from a carefully evolved network," Proceedings Robotics and Automation, Arizona, pp 692–694, (1989).
14. D. Cliff, I. Harvey, and P. Husbands, "Evolutionary robotics," IEE Colloquium on Design and Development of Autonomous Agents, London, UK, pp 1/1–1/3 (1995).

18

Robust Flight Stability and Control for Micro Air Vehicles

M. Meenakshi[1] and M. Seetharama Bhat[2]

[1] Dept. of Aerospace Engineering,
Indian Institute of Science Bangalore 560012, India.
meena@aero.iisc.ernet.in,
[2] msbdcl@aero.iisc.ernet.in

Summary. This chapter presents the real time validation of fixed order robust H_2 controller designed for the lateral stabilisation of a micro air vehicle named Sarika2. Digital signal processor (DSP) based onboard computer named flight instrumentation controller (FIC) is designed to operate under automatic or manual mode. FIC gathers data from multitude of sensors and is capable of closed loop control to enable autonomous flight. Fixed order lateral H_2 controller designed with the features such as incorporation of level 1 flying qualities, gust alleviation and noise rejection is coded on to the FIC. Challenging real time hardware in loop simulation (HILS) is done with dSPACE1104 RTI/RTW. Responses obtained from the HILS are compared with those obtained from the offline simulation. Finally, flight trials are conducted to demonstrate the satisfactory performance of the closed loop system. The generic design methodology developed is applicable to all classes of Mini and Micro air vehicles.

Keywords: Micro air vehicle, robust controller, flight instrumentation controller, hardware in loop simulation, autonomous flight.

18.1 Introduction

Recently, the desire for low cost, portable, low-altitude aerial surveillance has driven the development and testing of smaller and smaller aircrafts, known as micro air vehicles (MAVs). MAVs are envisioned as rapidly deployable, autonomous aerial robots equipped with a camera system. The ducted or shrouded rotary-wing MAV would peer through windows, over roofs or above tree canopies to support missions like search-and-rescue [1] and target acquisition and thus suited for hover-and-stare. Equipped with small video cameras and transmitters, fixed wing MAVs have great potential for surveillance and monitoring tasks in areas either too remote or dangerous to send human scouts. It should be noted that fixed-wing miniature aerial vehicles [2] typically fly around $20\,\mathrm{m\,s^{-1}}$ in open skies and thus not suited for hover-and-stare.

M. Meenakshi and M.S. Bhat: *Robust Flight Stability and Control for Micro Air Vehicles*, Studies in Computational Intelligence (SCI) **76**, 153–162 (2007)
www.springerlink.com

Fig. 18.1. Micro air vehicle- Sarika2

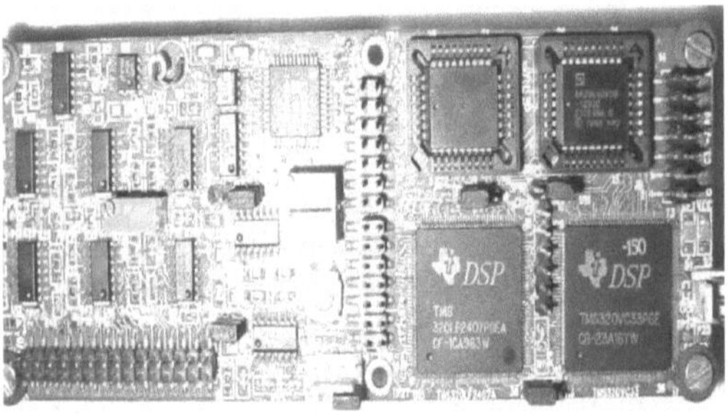

Fig. 18.2. Flight instrumentation controller

MAV's are radio controlled remotely piloted systems. These vehicles are difficult to fly due to their unconventional design and unpredictable flight characteristics. Another potential source of instability for MAV is the relative magnitudes of wind gusts, which are much higher at the MAV scale. Hence, to alleviate the necessity of an expert pilot, some form of augmented control must exist. Thus, in [3], a single lateral stability augmentation system (SAS), which meets the flying qualities, disturbance rejection and noise attenuation specifications for the entire flight envelope (flight speeds ranging from 16 to 26 $\mathrm{m\,s^{-1}}$) of a MAV named Sarika2 (Fig. 18.1) is designed. The controller is designed at the central operating point of Sarika2 (20 $\mathrm{ms^{-1}}$) and is used for its entire flight envelope. The software in simulation (SiL) results showed excellent performance over its entire flight envelope [3]. However, SiL does not validate the implementation of a controller. In a sampled time digital computer with real life I/O there is no guarantee that the real time performance constraints can be met [4]. The application of HILS technology to the controller design process can reduce expensive field tests, improve product quality, facilitate examination of subsystem interactions and resolve critical safety issues prior to in-vehicle testing [4]. This chapter presents the development and real time validation of DSP based FIC (Fig. 18.2) that gathers data from multitude of sensors and is capable of closed loop control to enable smooth flight of MAV. Simple augmented control is the initial step to enable autonomous flight.

The overview of this chapter is as follows. Sections 18.2 and 18.3, respectively, give a brief description of the vehicle and the design of fixed order robust H_2 controller. Offline simulation results are given in Sect. 18.4. Section 18.4 also gives the real time validation of fixed order robust H_2controller using the concept of HILS and the flight tests. Finally conclusions are drawn in Sect. 18.5.

18.2 Sarika2: Description

MAV named Sarika2 is designed at the Department of Aerospace Engineering, Indian Institute of Science, Bangalore, incorporates a clipped delta wing with the controls through the inboard aileron and outboard elevator on the wing and a rudder on the vertical fin. Sarika-2 carries a power plant, fuel, landing gear, radio receiver, batteries, servo controls for engine, rudder, elevon and sensors. The fuselage carries the power plant in the front and vertical fin and rudder at the back. It is like a box covered on three sides while the top is open. The high set wing largely covers the top after assembly. The devices needed for the flight of the Sarika-2 are all placed in the fuselage. The pertinent geometric and mass properties of the vehicle are: mass = 1.75 kg, wingspan = 0.6 m, wing-area = 0.195 m^2 moment of inertia of 0.043, 0.0686 and 0.07507 $\mathrm{k\,gm}^2$ along x, y and z axis, respectively.

18.2.1 Open Loop Analysis

Frequency and damping ratios of lateral dynamics of the aircraft are calculated at different operating points. The dutch roll frequency increases from 5.92 to 9.55 $\mathrm{rad\,s}^{-1}$ as the speed is varied from 16 to 26 $\mathrm{m\,s}^{-1}$. The dutch roll damping is very low, around 0.056 at 20 $\mathrm{m\,s}^{-1}$. The spiral time constant increases from 2.29 s to 3.048 s as speed changes from 16 to 26 $\mathrm{m\,s}^{-1}$. The roll subsidence is faster and its time constant decreases from 0.44 to 0.326 s as the speed increases from 16 to 26 $\mathrm{m\,s}^{-1}$. Though the lateral dynamics of Sarika2 is designed to be stable, there might be large variations of the stability derivatives during the flight. The fast dutch roll response with poor damping makes the flying difficult with radio control. Owing to its delta wing configuration, the vehicle has very small reaction time to external disturbances like gust, hence pilot correction is very difficult, if not impossible, during its open loop flight. Most critical issue is the technologies used in rate and acceleration sensors on larger aircraft are not currently available at the MAV scale. Since, Sarika2 does not have sensors to measure true or indicated airspeed and uses non-inertial quality sensors, gain/controller scheduling over the flight speed range of 16–26 $\mathrm{m\,s}^{-1}$ is not feasible. Hence, a single controller needs to be designed for robustness against variations in speed, weight, centre of gravity, time delays, non-linearities, and disturbances from wind gusts.

18.3 Design of Robust H_2 Controller

Reduced order robust H_2 controller is designed for the lateral stabilisation of Sarika-2. The controller is designed by solving strengthened discrete optimal projection equations, which includes a set of four coupled, modified Riccati and Lyaponov equations. It is found that a third order controller (given in (1) [3]) is sufficient to meet the level 1 flying qualities of Sarika2. Level 1 flying quality requirements are:(1).Dutch roll damping ratio ≥ 0.5, (2).Roll subsidence time constant <1 s 3). Spiral mode-minimum time to double the amplitude >12 s. Along with the level 1 flying qualities, the controller is also able to reject low frequency wind gust disturbance input and to attenuate high frequency sensor noise signal.

$$K(z) = \frac{1}{\Delta(z)} \begin{bmatrix} -0.68\,(z-1.07)\,(z-0.63) & -0.58\,(z-0.99)\,(z-0.61) & 0.17\,(z-1.66)\,(z-0.65) \\ -0.05\,(z+2.92)\,(z-0.94) & -0.02\,(z+5.24)\,(z-0.96) & 0.03\,(z+1.46)\,(z-1.04) \end{bmatrix} \tag{18.1}$$

where, $\Delta(z) = (z-0.72)(z^2-0.30z+0.29)$.

18.4 Validation of the Controller

The controller is validated by using offline and real time simulations. Finally the flight trials are conducted to demonstrate the satisfactory performance of the controller.

18.4.1 Robustness Analysis

Figure 18.3 shows singular values of the output sensitivity So, control sensitivity S_iK and the cosensitivity T of the closed loop system. Singular values of So for the lateral acceleration, roll rate and yaw rate show slight deviation from one another at low frequencies (below 8 rad s^{-1}) having a magnitude in the neighbourhood of 0 dB in the negative side. Lower values of S_iK at low frequencies (up to 40 rad s^{-1}) indicate very good stability robustness properties

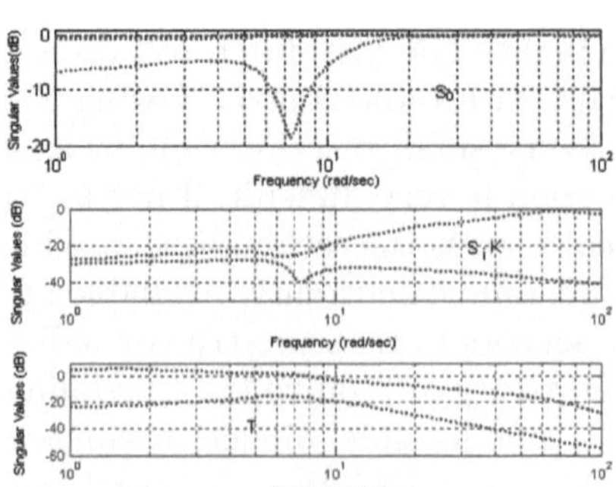

Fig. 18.3. Singular values of S_o, S_iK and T

against the additive uncertainty, Δa of the closed loop system. Though the upper bound of $S_i K$ increases above $40\,\text{rad}\,\text{s}^{-1}$, it is well within the 0 dB indicating positive gain margin. The singular values of T show an attenuation of −20 dB/decade above $\omega = 10\,\text{rad}\,\text{s}^{-1}$ assuring good noise attenuation by a factor of 10 which is acceptable.

18.4.2 Perturbation Analysis and Time Response Study

To evaluate the effectiveness of the controller in eliminating any errors caused by low frequency perturbations due to parameter variations, a 20% uncertainty on static derivatives and 50% uncertainty on dynamic derivatives are considered. Figure 18.4 gives the time responses of the perturbed plant and its closed loop system to a doublet input of magnitude ±0.1 m s (with respect to the neutral position of 1.6 m s) on the roll axis. A 20 m s delay is included at the input of aileron and rudder actuator blocks to take in to account computational delays. A wide band noise of zero mean and 0.005 V variance that represents the sensor noise is also introduced during the simulation. The aircraft is first trimmed for straight level flight at an altitude of 1,000 m above the sea level. Then a two second duration positive pulse of amplitude 0.1 m s (above the nominal value of 1.6 m s) corresponding to −3.84 ° aileron deflection is applied to the aileron channel at t = 1s. This input is followed by a 2s, negative pulse of amplitude 0.1 m s, which corresponds to a 3.84 ° aileron deflection. In Fig. 18.4, the responses up to 15 s belong to the un-perturbed (nominal) plant whereas for t >15 s are due to that of the perturbed plant. Though the perturbed plant is highly oscillatory, the closed loop responses of both perturbed and nominal plant settle down relatively faster than that of uncontrolled aircraft responses without any performance degradation. In addition, gust disturbance and sensor noise are rejected at all flight conditions. The fast settling and well-damped closed loop response demonstrates the robust performance against parameter variation.

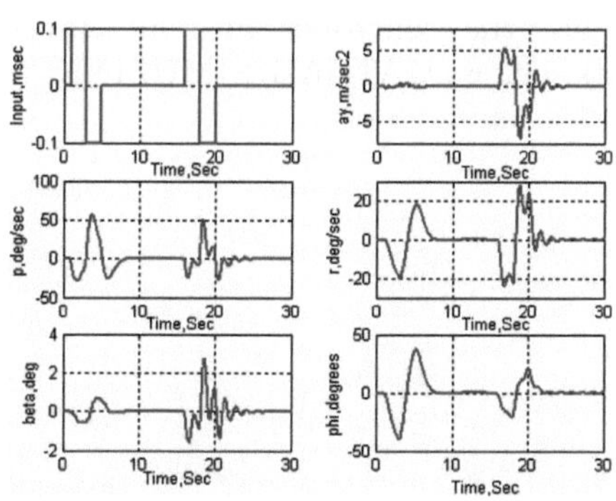

Fig. 18.4. Closed loop responses

18.4.3 Real Time Validation: HILS

The integration of a lightweight onboard computer, which is capable of performing closed loop control functions in real time, is a technically challenging task since off-the-shelf solution did not exist with the stringent constraint on size (limited space within Sarika2), weight and power. As a result, DSP (TMS320VC33 and TMS320LF2407) based FIC; Figure 18.2, (of size 139.7 mm × 50.8 mm and weight 55 gm) has been developed [5].

Floating point programming technique of DSP is used to develop all onboard software for the FIC. Main algorithm comprises of four main steps i.e.(1) initialisation routines. (2) delay routine.(3) sensor calibration routine, (4). interrupt enable routine which enables the timer interrupt, TINP, the capture interrupt CAP, the external interrupt 2, XINT2 and waits for the interrupt [5].

In order to validate that FIC can indeed meet real time performance constraints; the HILS technique is employed. In HILS, either a simulated plant is wired to a real controller or a real plant is wired to a simulated controller. In this work, a real controller is wired to the simulated aircraft as shown in the HILS experimental setup, Fig. 18.5. A HILS experimental setup includes a simulation computer, onboard FIC and ground station radio frequency control unit.

The feedback signals to the real time controller are determined by the lateral flight dynamic model, running on the desktop housing dSPACE1104 RTI/RTW. dSPACE1104 subsequently generates the equivalent analog signal in volts through its DAC unit. Sensitivities of the MEMS based sensors used for Sarika2 are: 1.11 mV/degree s^{-1} (with an offset of 2.25 V), Accelerometers: 500 mV g^{-1} (with a bias of 2.5V). The controller continuously gathers the sensor signals along with the possible pilot command inputs. Simultaneously, controller processes the received data and generates the actuator commands through its PWM generation units. The simulated flight dynamics model receives these outputs through its capture units allowing for assisted flight control.

In order to validate the controller in real time, powerful open loop and closed loop tests using MATLAB/SIMULINK tools and RTI/RTW

Fig. 18.5. HIL simulated plant, real controller

are employed. Under open loop test, controller receives the standard test inputs in time-indexed mode using computer scheduler utilities. In closed loop test, FIC receives the feedback signals through DAC/ADC interface of dSPACE1104/FIC. Finally satisfactory performance of the closed loop system is demonstrated by actual flight trials.

Figure 18.6 shows the inputs used for open loop test and Fig. 18.7 shows the corresponding output of the controller, recorded for a small interval of time. The input pattern is a pulse of amplitude 2 V and a duration of 8 s (from $t = 4$ s) on yaw rate channel followed by a doublet input of amplitude ± 0.1 V and a duration of 2 s (from $t = 12$ s) on lateral acceleration a_y channel, and a pulse of amplitude 0.2 V on roll rate channel (from $t = 21$ to 26 s.). For the comparison, Fig. 18.6 also shows the expected output of the controller obtained using SIMULINK tools. From the Fig. 18.7, one can see the excellent match between the actual output obtained from FIC and the expected output from the SIMULINK tools with a small offset of 0.04 m s (equivalent to 0.16° aileron deflection) on aileron channel. This small difference is due to the limitations of fixedpoint processor.

Figures 18.8a and b represent the real time command signal from radio frequency receiver, controller output and emulated sensor signals of the closed loop test corresponding to a flight speed of $20\,\mathrm{m\,s^{-1}}$.

Figure 18.8a includes the command input applied on the aileron channel by moving the aileron joystick and the corresponding inputs to the aileron and rudder actuators from the controller. The command input is around -0.1 to 0.1($\pm 3.84°$ degree aileron deflection) with respect to the neutral value

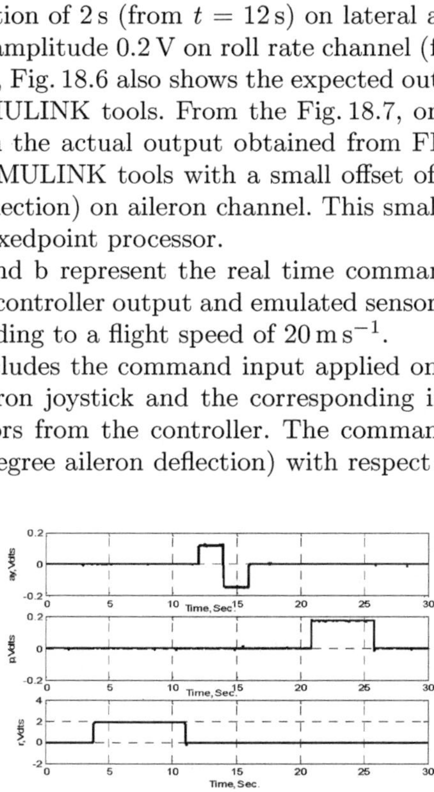

Fig. 18.6. Scheduler input sequence to SAS

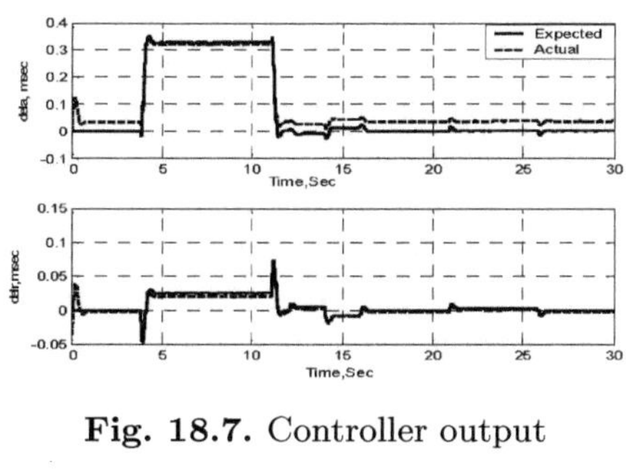

Fig. 18.7. Controller output

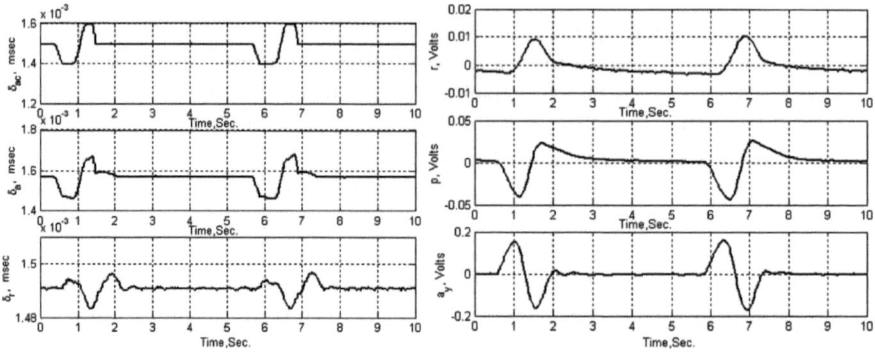

Fig. 18.8. (**a**) Command input and controller outputs. (**b**) Outputs of the emulated sensors

Fig. 18.9. Sarika-2 in flight

of 1.5 ms. In Fig. 18.8b the a_y response shows an excursion of a maximum of 0.15 V to -0.15 V. A 0.15 V corresponds to 0.3 g lateral acceleration. Similarly a -0.15 V corresponds to -0.3 g lateral acceleration, which is comparable with its offline simulation results. Simultaneously, the real time roll rate response ranges from -0.04 to 0.02 V with respect to the neutral position. This corresponds to a roll rate of -36 degree s^{-1} with a -0.1 ms command input. Corresponding offline simulation shows a roll rate of -29.6 degrees s^{-1}. Hence, real time simulation responses match well with a small difference compared to that of offline simulation responses. Similarly, the open loop and closed loop HILS is repeated at different flight speeds like, 16 and 26 m s^{-1}. The responses are found to be satisfactory.

Next, Sarika-2 is flown (Fig. 18.9) in traditional manual or radio controlled mode without flight stabilisation for further data analysis.

The sensors output (Figs. 18.10 and 18.11) recorded during taxiing run with flight stabilisation and flight would indicate quality of signals from low cost MEMS rate gyros and accelerometer, impact of engine vibration on sensors output, ground reaction on the vehicle during the taxiing and would give confidence on the plant model used for the controller design. Figure 18.12

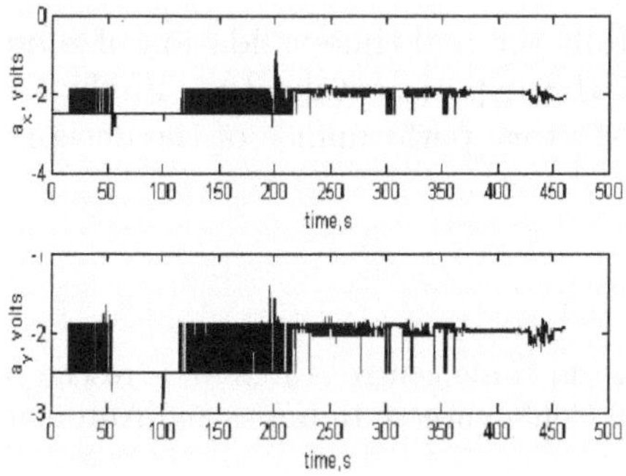

Fig. 18.10. a_x, a_y, responses from flight trial

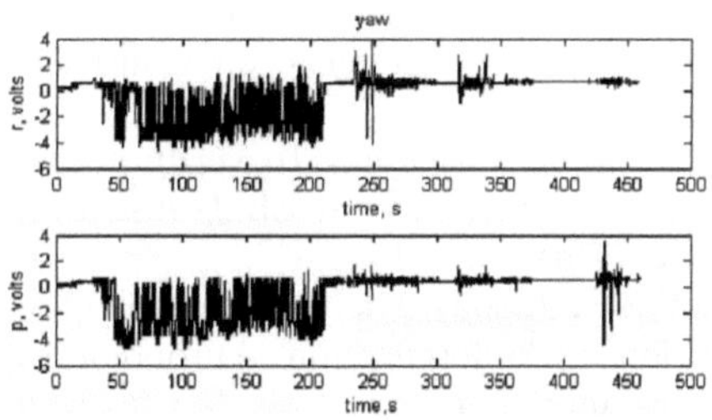

Fig. 18.11. p, r responses from flight trial

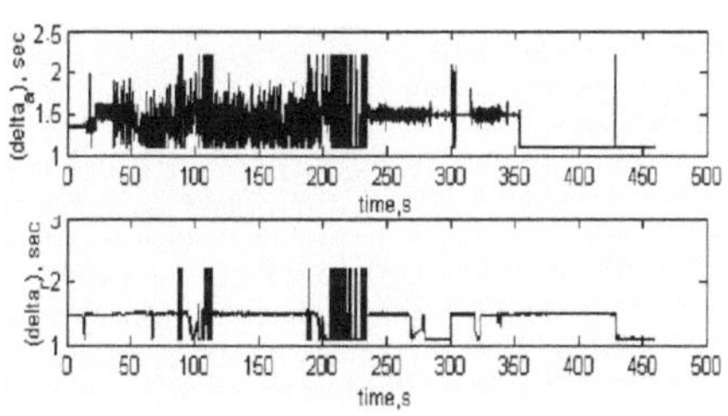

Fig. 18.12. Controller outputs

shows the controller outputs generated during the test trials. Analysis of the controlled flight proved that the aircraft remains stable under closed loop with improved damping. Moreover, the pilot on ground is satisfied with the performance of the controller during the test trials, which proves that the controller works satisfactorily.

18.5 Conclusion

This chapter explains the way to enable autonomous flight of the MAV through the inclusion of onboard flight instrumentation computer, capable of using rate gyros and accelerometers to assist in the control of the aerial vehicle.

This chapter also details the real time validation of robust fixed order lateral H_2 controller using HILS experimental setup. Flight trials are conducted to demonstrate the satisfactory performance of the designed controller.

References

1. Blitch J, (2002) World trade centre search-and- rescue robots, Plenary Session IEEE International Conference on Robotics and Automation, Washington D.C.: 319–325.
2. Grasmeyer JM, Keennon MT, (2001) Development of the Black Widow micro air vehicle, 39th AIAA Aerospace Sciences Meeting and Exhibit, Reno, NV: 1–9.
3. Meenakshi M, Bhat MS, (2006) Robust fixed order lateral H_2 controller for micro air vehicle, 9th International Conference on Control, Automation, Robotics. ICARCV2006, Singapore.
4. John RW, John FK,(1997) A strategy to verify chassis controller software—dynamics, hardware, and automation, IEEE Transactions on Systems, Man and Cybernetics - Part A, 27(4) : 480–493.
5. Meenakshi M, Bhat MS, (2006) Micro aerial vehicles- Towards autonomous flight, 3rd International Conference on Autonomous, Robots and Agents, ICARA2006, New Zealand : 7–12.

19

An Ecological Interface Design Approach to Human Supervision of a Robot Team

H. Furukawa

Department of Risk Engineering,
University of Tsukuba, Tsukuba, Japan
furukawa@risk.tsukuba.ac.jp

Summary. When supervising a team of robots, the operator's task is not only the manipulation of each robot but also achievement of the top goal that is assigned to the entire team of humans and robots. A main goal of this study is development of a design concept based on the ecological interface design for human supervision of a robot team, providing information about states of functions that are necessary to understand the overall progress in the work and the situation. This paper describes an experimental study conducted to reveal basic efficacy using an experimental test-bed simulation. The results suggest that the proposed approach has the ability to enable effective human supervision.

Keywords: Human–robot interface, Ecological interface design, Functional displays, System-centred view, Multiple robot management.

19.1 Introduction

Regardless of use of multiple robots, the operator's task involves not only manipulation of each robot but also achievement of the top goal that has been assigned to the entire team of humans and robots. There are several factors which pose a challenging problem in supervision and management of multiple robots. Although cognitive resources of humans are limited, operators are demanded to understand highly complex states and make appropriate decisions in dynamic environment. Furthermore, operators have to deal with large amounts of complex information which may risk overwhelming them in the supervision tasks. As a consequence, there has been increased interest in developing human–robot interfaces (HRIs) for human supervision of multiple robots [3].

The main goal of our project is the development of an interface design concept based on *ecological interface design* (EID) for human supervision of a robot team. EID is a design approach based on the externalization of the operator's mental model of the system onto the interface to reduce the

H. Furukawa: *An Ecological Interface Design Approach to Human Supervision of a Robot Team*, Studies in Computational Intelligence (SCI) **76**, 163–170 (2007)
www.springerlink.com

cognitive workload during state comprehension [9, 10]. EID provides information about states of functions that are necessary to achieve the top goal of a human–machine system. Information on function is identified using the abstraction–decomposition space (ADS). An ADS is a framework for representing the functional structures of work in a human–machine system that describes hierarchical relationships between the top goal and physical components with multiple viewpoints, such as abstraction and aggregation [7]. Since the operator's comprehension of the functional states based on the ADS is an essential view for the work, supporting the view is crucial for them to make operational plans and execute the plans appropriately under high-workload conditions [5]. However several attempts have been made to apply the design concept to HRI [4, 8], empirical evidence of the effectiveness of this approach, while necessary, is not sufficient.

This paper describes an experimental study conducted to reveal the basic efficacy of EID in human–robot interactions, where the material was first presented at [2]. The next section explains an application and implementation approach to a test-bed simulation platform of the proposed design concept, and a procedure of the experiment using the prototype system. We then discuss the results to examine the usefulness of ADS for representing whole tasks allocated to humans and robots, the feasibility of designing indications for the functions in the ADS, and the efficacy of function-based interface design to improve human–robot collaboration.

19.2 Methods

19.2.1 The RoboFlag Simulation Platform

This study uses the RoboFlag simulation, which is an experimental test-bed modeled on real robotic hardware [1]. The chief goal of an operator' job is to take flags using home robots and to return to the home zone faster than the opponents. The basic tasks for achieving this goal are two: *Offence* and *Defence*. One operator directs a team of robots to enter an opponent's territory, capture the flag, and return to their home zone without losing the flag. Defensive action takes the following form: a rival robot will be inactivated if it is hit by a friendly robot while in friendly territory.

Figure 19.1 shows the display used for an operator to monitor and control his or her own team of robots. A circle around a robot indicates the detection range within which the robot can detect opponents and obstacles. The simulation provides two types of operations that operators can select according to their situation: manual controls and automatic controls. In manual control mode, an operator indicates a waypoint to a robot by clicking the point on a display. Two types of automatic controls were implemented in this study. When *Rush and Back* (R&B) mode is assigned, the robot tries to reach the

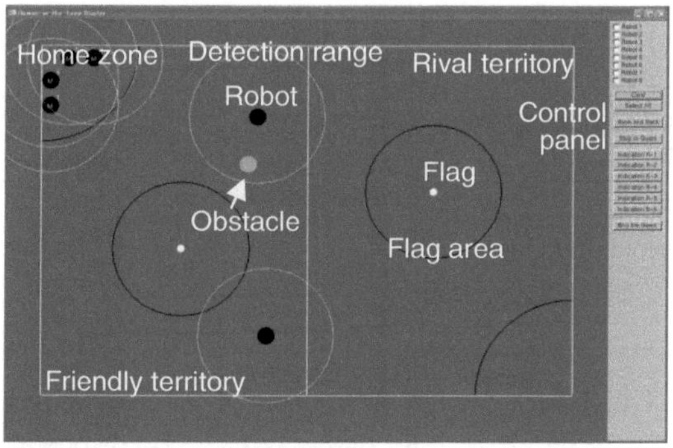

Fig. 19.1. Display for the RoboFlag simulation

flag and returns home after it captures the flag. The course selected is straightforward, in that the robot heads directly to the destination. In *Stop or Guard* (S/G) mode, the robot stays in the home position until it detects an opponent. If an opponent robot comes into detection range, the robot tries to inactivate the opponent.

The robots are semi-autonomous, that is to say, they have the ability to change their own courses to avoid rival robots or obstacles. Offensive and defensive tasks of the rival robots were fully automated by using the two types of automatic controls implemented in this study.

Because time constraints are severe in this RoboFlag game, human operators need gain an understanding of the situation as rapidly as possible. Furthermore, it is necessary for operators to comprehend the state of entire area as well as the local area.

19.2.2 Implementation of Functional Indications

The following are descriptions of two function-based interface designs, in which one was designed to represent the state of a lower-level function under an *Offence* function, and the other under a *Defence* function. Previous studies using the RoboFlag simulation showed that human–robot interactions depend on various contributory factors [6]. Because of the high complexity of the interactions, a part of the whole ADS was selected as the target functions indicated on the interface in this study.

To specify each state of the function, expressions that graphically showed the state in the physical relations between each robot and the object was used. This has aimed to enable operators intuitive to understand the state of the functions.

Figure 19.2 depicts a part of the ADS whose top is the *Offence* function. One of the means of achieving the function *Avoid opponents* is *Set way-point such as not to encounter opponents.* To select an appropriate course to reach the flag, the situation along the course, especially the positions of opponents,

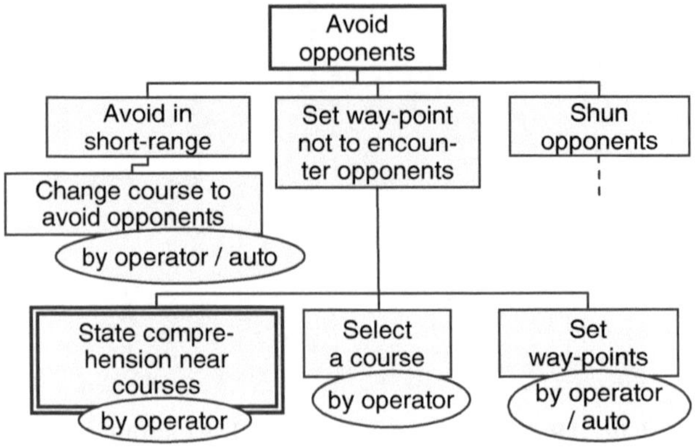

Fig. 19.2. An ADS for the function *Avoid opponents*

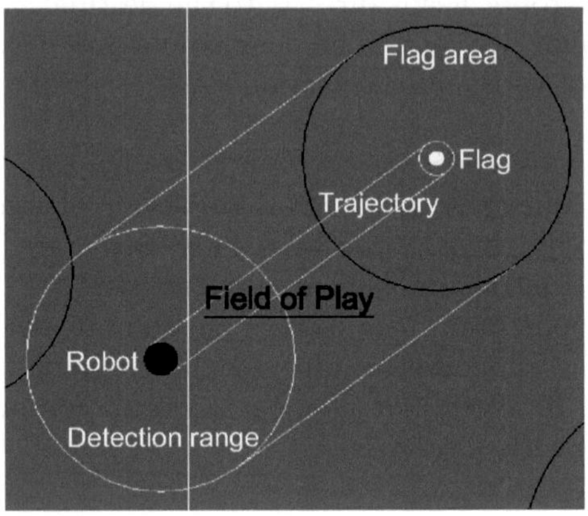

Fig. 19.3a. An interface design indicating the state of the function *State comprehension near courses*

should be understood by the decision-maker. The proposed indication was applied to the function *State comprehension near courses*, which is one of the key sub-functions included in the *Offence* function, and is allocated to the human operator. The indication is depicted in Fig. 19.3a. A robot is shown as a black circle and the flag as a white circle. The two straight lines connecting the robot and the flag show the trajectories along which the robot is going to move. The two lines on the outside, which connect the detection range and the flag area, show the range in which detection becomes possible when the robot moves along the route. In other words, opponents in this area can tackle their own robots moving along the course. The display clearly indicates the *Field of play* of the target task. One of the operator's options is to send a robot as a scout to the field if there is an area where the situation is unknown.

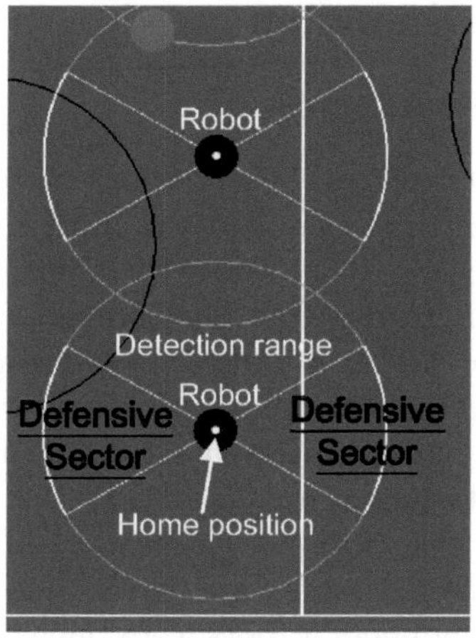

Fig. 19.3b. An interface design indicating the state of the function *Cooperation between defensive robots*

Cooperation between defensive robots is a type of defensive function realized by a team of robots, and an indispensable sub-function for achieving the *Defence* function. The picture illustrated in Fig. 19.3b is the functional indication designed for enabling an operator to be clearly aware of the state of the function. A circle around a robot indicates the detection range as described in the previous section. A fan-shaped sector, a *Defensive sector* is where a robot in S/G mode has a high ability to intercept opponent robots coming through. Outside the Defensive sector, the possibility of catching opponents is lower than within the sector. An operator can use spaces between the sectors as an indication of the defensive ability of the defensive robot team in the position.

19.2.3 Procedure

Twenty-two paid participants took part in the experiment. The participants were randomly divided into two groups of eleven. One group (the *original* group) used the original human-robot interface for the RoboFlag simulation, and the other group (the *modified* group) used the modified interface display designed according to the proposed concept.

The participants learned their tasks, rules of the game and the details of the assigned HRI, and mastered skills for controlling the team of robots through playing the game. They were asked to try it out until they found their own strategies to play the game. After they had decided on their strategies, they played the game five times as part of the main experiment. At the end of each game, they were asked to write the details of their strategies and usage of information represented on the display. The quantified data acquired in the main experiments were then statistically analyzed.

19.3 Results and Discussion

19.3.1 Statistical Data Analysis

The number of flags captured was counted for every game. The averages and standard deviations of participants' captures in the *original* and *modified* conditions are M = 0.75, SD = 0.62, and M = 1.20, SD = 0.85. A repeated-measures ANOVA test indicates that the difference between two conditions is significant ($F(1, 20) = 6.164$, $p < 0.05$). This result may suggest that the modified display is effective in supporting operators in their offensive task, regardless of their ability or the strategy used for the task.

The results of the statistical analysis show that there are no significant differences between the *original* and *modified* conditions for the number of flags captured by opponent robots, win percentages, the numbers of times that participants' and opponents' robots were tagged, total elapsed times, and time before the first capture by participants' and opponents' robots. However, at least, the results show no sign of any ill effects caused by using the modified interface.

19.3.2 Strategies Developed and Use of Functional Indications

This section illustrates the strategies developed by the eleven participants of the *modified* group and how they used the information on functions represented on the display.

For offensive operations, five participants mainly used the R&B automatic operation to capture the flag. Four of them, i.e. 80% of the five operators, tried to comprehend the state of the robots and situation around the course using the *Field of play* indication. For defensive operations, ten participants allocated two to four robots on a course that opponent robots followed to capture the flag. Eight of ten, i.e. 80% of them, used the *Defensive sector* indication to decide appropriate spaces between the guarding robots at the training phase and/or the main experiments. Their usage and target functions exactly match with those expected in designing phase.

The six participants who chose manual controls for offensive actions fixed all the waypoints and timings of the orders in advance. The *Field of play* indication was not necessary for them during the main experiments. In spite of this, they mentioned that the indication was useful for developing their own strategies during the trial-and-error processes in the training. One participant who used the manual-controlled strategy for offence decided not to take any defensive action. A swift attack was his only strategy. The *Defensive sector* display is not necessary for this strategy.

19.3.3 Discussion

The analysis on the operators' uses of the functional indications suggests that definition of functions specified in the ADS meets the participants' understanding of functions, and that the ADS includes all the functions to which participants directed their attention in the operations. It also demonstrates that the functional indications, which are designed for the functions, were useful for participants to comprehend states of the functions.

The results also indicate that the need for a functional display closely depends on the strategies actually used during operations. This result suggests that individual difference in strategies should be taken into account when designing suitable interface displays for supervising multiple robots.

As for this experiment, the functional indications added to the original display did not cause obvious harm to the participants even when the information was not necessary in their operations. It can be said that the ADS and the interface display based on the ADS were appropriately built, which do not cause any interference in participants' supervision.

These findings may lead to the conclusion that the proposed design concept can offer a proper framework for developing HRIs which provide effective human supervision of multiple robots.

19.4 Conclusion

This paper describes an experimental study conducted to reveal basic efficacy of a human–robot interface design concept, in which the *ecological interface display* approach is used as the basic framework for implementing the information about a human–robot team work into an interface display. The results may suggest that the whole work can be modeled using ADS, and it is feasible to design useful functional indications based on the ADS. This study provides empirical evidence for the efficacy of the proposed approach to enable effective human supervision of multiple robots.

The results also show the need to consider two factors to design effective HRI displays: the one is participants' strategies developed for tasks, and the other is how they use the functional indications.

To elaborate the practical and effective design concept for HRIs, several techniques are necessary. Typical examples are methods for designing functional models for target tasks using an ADS as a framework, methods for selecting functions for which support of comprehension is necessary for operators, and methods for designing effective indications for easy understanding of states of the functions.

19.5 Acknowledgements

This work has been supported by Grant-Aid for Science Research (C) 18500104 of the Japanese Ministry of Education, Science, Sports and Culture.

References

1. Campbell M, D'Andrea R, Schneider D, Chaudhry A, Waydo S, Sullivan J, Veverka J, Klochko A (2003) RoboFlag games using systems based hierarchical control. In: Proc. the American Control Conference, Denver, USA, pp 661–666
2. Furukawa H (2006) Toward ecological interface design for human supervision of a robot team. In: Proc. 3rd International Conference on Autonomous Robots and Agents - ICARA'2006, Palmerston North, New Zealand, pp 569–574
3. Goodrich MA, Quigley M, Cosenzo K (2005) Task switching and multi-robot teams. In: Parker LE, Schneider FE, Schultz AC (eds) Multi-robot systems: From swarms to intelligent automata, vol. III, Springer, Netherlands, pp 185–195
4. Jin J, Rothrock L (2005) A visualization framework for bounding physical activities: Toward a quantification of gibsonian-based fields. In: Proc. the Human Factors and Ergonomics Society 49th Annual Meeting, Orlando, USA, pp 397–401
5. Miller CA, Vicente KJ (2001) Comparison of display requirements generated via hierarchical task and abstraction-decomposition space analysis techniques. International Journal of Cognitive Ergonomics 5:335–355
6. Parasuraman R, Galster S, Squire P, Furukawa H, Miller C (2005) A flexible delegation-type interface enhances system performance in human supervision of multiple robots: Empirical studies with RoboFlag. IEEE Transactions on Systems, Man, and Cybernetics - Part A 35:481–493
7. Rasmussen J (1986) Information processing and human-machine interaction. Elsevier Science Publishing, New York
8. Sawaragi T, Shiose T, Akashi G (2000) Foundations for designing an ecological interface for mobile robot teleoperation. Robotics and Autonomous Systems 31:193–207
9. Vicente KJ (2002) Ecological interface design: Progress and challenges. Human Factors 44:62–78
10. Vicente KJ, Rasmussen J (1992) Ecological interface design: Theoretical foundations. IEEE Transactions on Systems, Man, and Cybernetics 22:589–606

Embedded RTOS: Performance Analysis With High Precision Counters

Kemal Köker

Department of Computer Science
Computer Networks and Communication Systems
University of Erlangen-Nuremberg, Germany
koeker@informatik.uni-erlangen.de

Summary. We evaluated the performance of an embedded real-time system on a PC/104 system for the usage in a scenario of soccer playing robots. For this purpose we patched a common Linux kernel with the real-time application interface RTAI and installed it on the robots' embedded systems. We performed a performance analysis of the operating system by monitoring the response time for externally caused hardware interrupts to verify the usability of connected IR distance sensors to the on-board embedded system. Therefore we implemented an interrupt service routine for the board's parallel port to generate a system response for externally triggered hardware interrupts. For a faster recording interrupts were triggered via a signal generator connected to the systems' parallel port and monitored the response via an oscilloscope. Because of errors for higher frequency inputs and for detailed statistical analyses, we built a monitoring system by using a high-precision histogram scaler and counter. The performance of the embedded system with respect to the response time has been monitored and analysed in various system states. Our infrastructure allows an easy and precise possibility for performance analysis of embedded real-time operating systems.

Keywords: embedded Linux systems, real-time operating system, performance analysis

20.1 Introduction and Overview

Simulations are frequently used to test the usability of systems in different operating scenarios and are usually based on a system model that focuses the relevant parts. The analysis of the usability also includes a performance analysis of the operating systems which requires e.g. monitoring the response time for interrupts that are triggered by an input (e.g. pushing a button or getting new values of sensors). Usually the response time depends on additionally running tasks in the system and the scheduling algorithm and varies with respect to the system state (load or idle). Real-time operating systems guarantee a maximum time for the systems response for

K. Köker: *Embedded RTOS: Performance Analysis With High Precision Counters*, Studies in Computational Intelligence (SCI) **76**, 171–179 (2007)
www.springerlink.com

interrupt and represent a quite interesting possibility to be used in soccer robots. Smart scheduling algorithms and other techniques are used to keep the response time as low as possible. Mostly these techniques are distinguished for an one-shot or periodic occurrence of interrupts in a time period. In case of an one-shot interrupt, the operating system has to response as quickly as possible without any precautions for further interrupts. In case of a periodic mode, precautions for the scheduling are made to keep the response time lower. Considering the overhead for a one-shot mode, there should be a significant difference to the periodic-mode when the system has to response to higher interrupt frequencies. Therefore the mean of the response time in one-shot mode should lead to a higher value than the mean of the periodic mode. We build up a setting to confirm the upper conclusion also for embedded real-time operating systems. To keep it cost efficient and as easy as possible, we patched a Linux kernel with the real-time application interface RTAI [10] and installed a run capable embedded Linux operating system on a compact flash card for to use it in a PC/104 embedded system [4]. We used the PC/104 module from Arbor [5] with an AMD Ultra Low Power Geode GX1-300-MHz fan less CPU and on-board compact flash socket. This embedded board is similar to an off-the-shelf desktop computer with respect to its interfaces (serial-, parallel-, LAN-, USB-ports).

20.2 Embedded Linux System

The embedded operating system for our test setting was composed of the Linux kernel version 2.6.9 [6], BusyBox [7] for the system utilities, and uClibc [8] for the C library of the system. BusyBox combines tiny versions of common Linux/UNIX utilities into a single small executable (including the Web server). For this reason, it is also called a multi-call binary combining many common UNIX tools. UClibc is an optimized C library for the development of embedded systems. It is much smaller than the common GNU C library, but nearly all applications supported by glibc work without problems with uClibc. One more benefit is that uClibc even supports shared libraries and threading. Additionally all robots allow remote logins to the running system via a tiny SSH client/server from [9] for monitoring running processes. The operating system is stored on a compact flash card and because of the limited write-cycles, the system is running in ramdisk mode by loading a compressed initial ramdisk at booting. Table 20.1 gives a brief overview of the storage size for our fully functionally embedded Linux system.

Table 20.1. Component size of embedded Linux

Component	Name	Uncompressed	Compressed
Grub loader	grub	–	200 kB
Kernel 2.6.9	bzImage	–	1 100 kB
Root FS	initrd	4,400 KByte	2 100 kB
		Total	3 400 kB

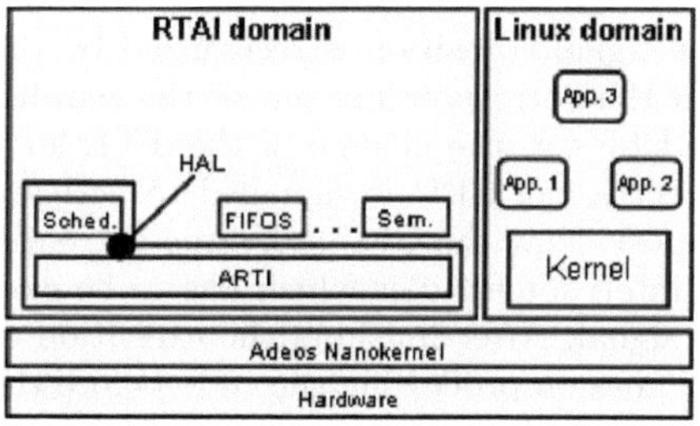

Fig. 20.1. Software architecture of RTAI

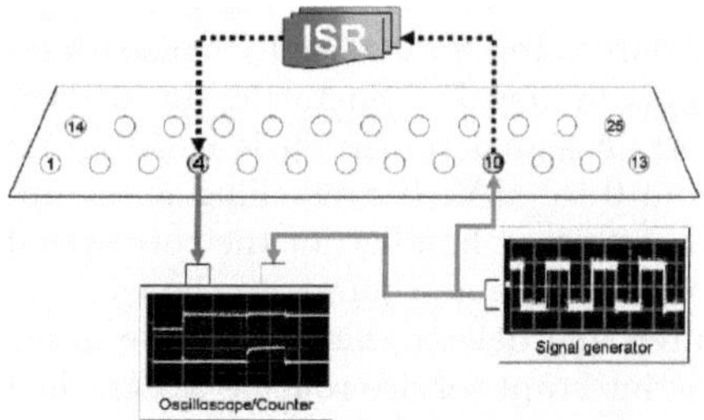

Fig. 20.2. Measurement infrastructure using oscilloscope

20.3 Real-Time Application Interface on PC/104

We looked for a free and open source real time version of Linux to keep the project cost efficient and selected RTAI. The well known real-time application interface RTAI is an extension and modification of a common Linux kernel. For the usage, the Linux kernel has to be patched with RTAI, which enables the operating system to response in a fast and predictable way. Fast means that it has low latency, thus i.e. it responds to external, asynchronous events in a short time. Predictable means in that case that it is able to determine tasks completion time with certainty. The current architecture of an RTAI system consists of the ADEOS Nanokernel layer (Adaptive Domain Environment for Operating System) [12], the hardware abstraction layer HAL and further modules for the requested functionality. The ADEOS Nanokernel layer is inserted as a division between the lower hardware and the operating system and has the task to transmit interrupts from the lower layer to the upper layer. Using ADEOS enables different operating systems to run at the same time, in which each of the operating systems registers its own handler to process an interrupt. Each operating system is assigned to a domain with a certain priority. Depending on the priority level, ADEOS manages the transmission of the interrupt to the corresponding operating system, where the higher priorities are provided first and the lower priorities get the interrupt signal at last. According to the architecture of ADEOS, there are at least two operating systems needed to make RTAI run. The first domain in our test system is RTAI as the real-time domain with a higher priority, and the second domain is the normal Linux kernel as the non-real-time part with a lower priority. In case of an interrupt, ADEOS toggles the interrupt first to RTAI because of the higher priority and thereafter to the lower priority Linux kernel. The HAL again consists of an ARTI layer (ADEOS based Real Time Interface) and the real time scheduler. The HAL enables an easy access of the OS to the interrupts via ARTI. Figure 20.1 shows the schematic architecture of an RTAI system with the Linux domain as the nonrealtime part.

20.4 Preparing for Performance Analyses

Performance evaluation of a system includes e.g. to monitor and measure response times for interrupt latencies. An interrupt may be generated either via software or hardware. An externally caused hardware signal transition is recognized by the Linux system by generating an interrupt on the corresponding pin of the parallel port. A signal transition is usually indicated by e.g. the change of the TTL level from 0 to +5V. If everything is set up correctly, the CPU is signalled to activate the interrupt handler for the corresponding port. Now the CPU stops the currently running code and jumps to the code of the interrupt handler which has to be executed und deletes the flag for the interrupt signal. After finishing the execution of the interrupt service routine (ISR), the CPU goes on processing the code before the interrupt occurred. For the analysis we implemented an ISR for the PC/104s parallel port that generates an echo for each input pulse and recorded the delays. Our implementation is using the RTAI API [11] and consists of three functions, called xinit_module, handler and xcleanup_module. The function xinit_module is called up automatically at module loading and is initialising the ISR and registering an interrupt handler for the interrupt 7 (parallel port). In case of an interrupt, a specific RTAI function is executed and a timer is started for the one-shot or periodic mode. The timer mode is defined by the RTAI module that is loaded at system booting. The handler function is the main function of the interrupt handler and generates just a short high signal and thereafter signals RTAI the end of service and standby for the next interrupt. The end of the service is indicated by setting a low level on the pins 2–9. The xcleanup_module is called up automatically on unloading the module and cleans up the interrupt for the parallel port. Our performance analysis for the operating system focuses on the delay between the signal transition and the corresponding response for it and is a modified version of the consideration in [2]. Equation 20.1 represents the basic considerations of [1] and characterizes the delay d_{echo} that is composed of the hardware propagation delay for an incoming signal pulse d_{hwi}, the interrupt latency d_{lat}, a delay to process the ISR d_{ISR} and the hardware propagation delay for the outgoing echo signal d_{hwo}.

$$d_{echo} = d_{hwi} + d_{lat} + d_{ISR} + d_{hwo} \tag{20.1}$$

While the other delays remain more or less constant and can be compensated for, d_{lat} depends on the system state at the time of the signal reception. The generation of the echo signal inside the interrupt service routine makes the assumption of a constant d_{ISR} reasonable because other interrupts are disabled for the processing time of the ISR. With respect to the start of recording where $t(n)$ is the time for the generation of the n-th signal pulse, the recording time for this event is (equation 20.2)

$$t_{rec}(n) = t(n) + d_{hwi}(n) + d_{lat}(n) \tag{20.2}$$

and the time when the echo pulse is observable as an external response signal transition is given in equation 20.3 as

$$t_{echo}(n) = t(n) + d_{echo} \Leftrightarrow t_{echo}(n) = t(n) + d_{hwi} + d_{lat} + d_{ISR} + d_{hwo} \tag{20.3}$$

Based on the above characteristics, the ISR for the parallel port allows recordings for the system response for signal transitions and further analyses for the performance.

20.5 Monitoring the Latency Using an Oscilloscope

To verify the proper function of our ISR we performed some measurements by using a signal generator and a digital sampling oscilloscope. We modified the test bed in [3] and used the frequency generator GF266 from ELC [13] to generate edges at TTL level and the 4-channel digital oscilloscope WaveSurfer424 from [14] to monitor and visualise the system response. Therefore the output of the signal generator is connected to pin 10 of the systems parallel port to cause interrupts and parallel connected to channel 1 of the digital oscilloscope. Each time a high level is on pin 10 of the parallel port, the ISR is processed and the corresponding response is served on e.g. pin 4 which is connected to channel 2 of the oscilloscope. The previous figure 20.2 displays the measurement infrastructure using the digital oscilloscope. The first test run was done at a signal rate of 10 kHz and an idle state of the embedded RTAI Linux system. The response time for an interrupt was not influenced by any other factors than the running task of operating system. Figure 20.3 displays a screenshot of the oscilloscope with a latency of 4.3 µs for the response time at 10 kHz. Even tough the system was in idle state, we monitored some fluctuations for the response time. To visualise this fluctuations, we used the cumulative sampling function of the oscilloscope, whereby the sampling is recorded continuously and additionally overdrawn about the previous signal course. The frequency of the occurrence is now represented by coloured pixels, in which a red colour symbolizes a higher and the violet colours a lower appearance of the recorded value (Fig. 20.4). Accounting the oscilloscopes recordings, we observed response times in the range of about 4 to 17 µs when the system is in idle state. We increased the frequency of the rectangle signal up to 100 kHz to make a simple check for the performance of the embedded system. The monitored data was almost equal to the first run and the increased interrupt frequency did not affect the response time. To evaluate the system performance, the system load usually should be increased by executing parallel programs. We started a second run and executed just the ping flood to the embedded Linux system via a 100MBit LAN connection and an interrupt frequency of 10 kHz. In this configuration, the signal course changed significantly considering the cumulative recording of the system response (Fig. 5). We monitored a response time in the range of 4 to 35 µs and peeks up to 40 µs. Even though the interrupt frequency and the system load increased, the response was still in time. Still in time means that we could

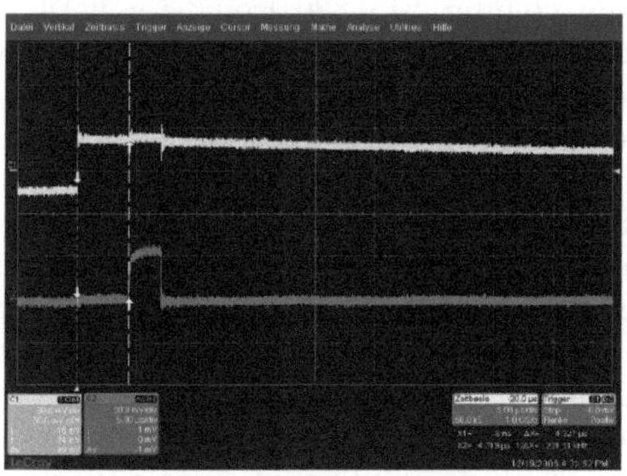

Fig. 20.3. Screenshot of the oscilloscope - latency for response 4.3 µs

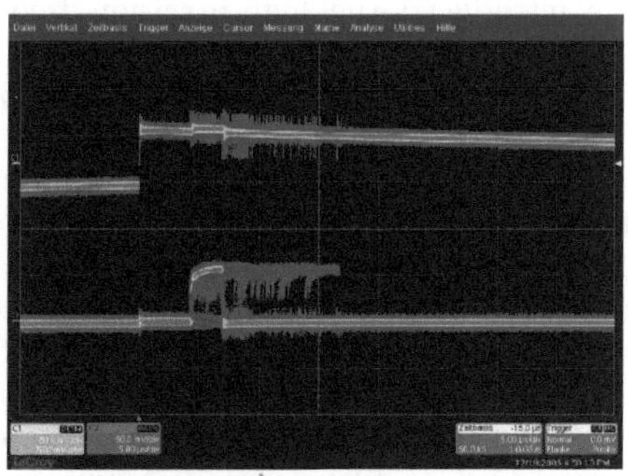

Fig. 20.4. Cumulative recording for the response time - system: idle

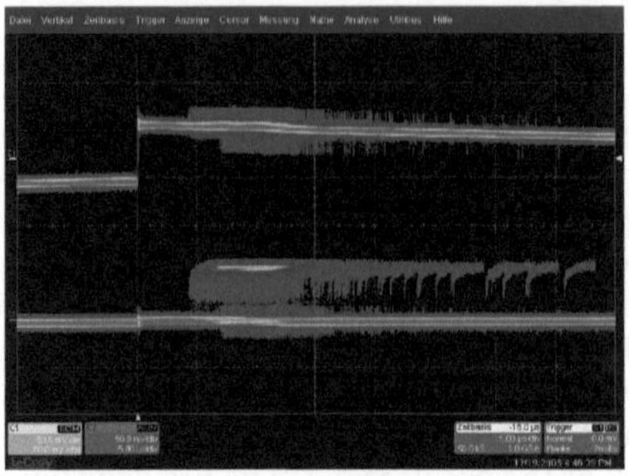

Fig. 20.5. Response time at 10 kHz interrupt frequency - state: load

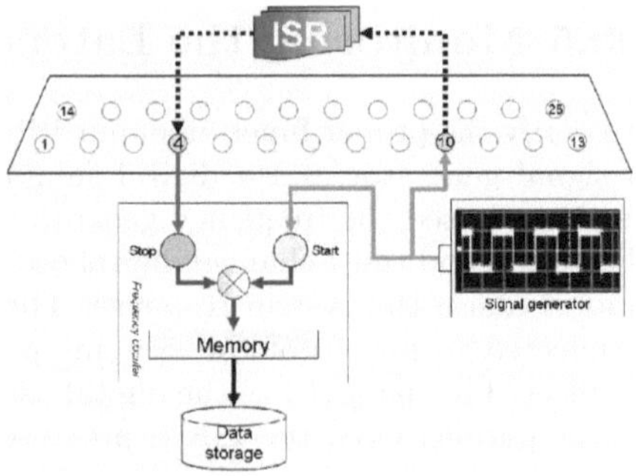

Fig. 20.6. Schematic of the measurement via frequency counter

relate each interrupt signal to its corresponding system response and vice versa. This bijectivity between interrupt and response is important for the evaluation and statistical analyses e.g. histograms. We started a third run and raised the interrupt frequency slowly up to 100 kHz. The system was loaded again by execution of a ping flood via LAN from an external computer. In this configuration (frequency 100 kHz) it was not possible to relate the interrupt signals to the corresponding responses. We monitored the beginning of the lagged responses at about 83.5 kHz and also some mavericks before. All mentioned variations of the test run are not useful for statistical analyses of the performance evaluation. However, our lab construction allows a first overview for the system performance of our embedded Linux system and externally caused interrupts and represents just a screenshot of the monitoring time. Detailed analyses of the performance require to record time related data for our lab construction for a long time period. The recorded data should contain the difference of the interrupt and the corresponding response time for each generated interrupt signal.

20.6 Measurement Infrastructure of Frequency Counter

We introduced the high-precision frequency counter SIS3820 from [15] with an internal 50MHz clock and a resolution of 20ns to gather time-related data. The frequency counter has a start and a stop port to initiate and terminate the measuring. According to the monitoring via an oscilloscope, we connected the start port of the frequency counter with the output signal of the frequency generator and to pin 10 of the parallel port. The stop port of the counter is connected to the response pin 4 of the parallel port. The duration between a input pulse and the response is measured via the counter by counting the clock ticks between the start and stop signal. Figure 20.6 displays the schematic for the measurement. To perform statistical analyses we need to collect the pair of interrupt signal and time difference for that interrupt and its response. Reliable statements about the system performance need a recording for a long period of time or huge amounts of interrupt signals triggered from the frequency generator. The frequency counter of the test bed allows to store up to more then 16 million values in a 64MByte memory in FIFO manner. For longer duration, a fast transmission of data to an external storage is required. We used our construction to record the response delay for 3.6 million interrupts in four different running modes. The test duration lasts 6 minutes at an interrupt frequency of 10 kHz. The tests differ for the one-shot or periodic scheduling and whether the system state is idle or load. For a better comparison of the results, we again increased the system load by executing a ping flood via LAN from an external computer and recorded the response times for the interrupts.

20.7 Statistical Analyses of Example Measurements

To evaluate the system performance based on statistical data, we took a closer look to the quartiles of the response times. The difference of the first and third quartile is defined as the distance of the quartile QD. This distance should cover 50% of all values and is used as the mass of the distribution. Usually this mass is robust against runaways and is usable for statistical evaluations. The left part of Table 20.2 gives a brief overview of the collected data for the response time using the frequency counter. Comparing the response times for the quartile distance delivers no significant difference between both scheduling modes for the idle state of the system. We just get a very small difference of 0.1 µs whereby the response time for the one-shot mode is faster. Increasing the system load also increases the quartiles for the response time. Comparing the results with the idle state, we notice a significant difference for the quartile distance and a faster response time for the periodic mode. This difference may result from the scheduler in the periodic mode, in which the scheduler is periodically executed. This means that the scheduler is prepared for further interrupts and the system can response faster as in one-shot mode. The results and the analyses of the quartiles are unique but for robust statements for the system performance we considered the statistical evaluation like minimum, maximum, mean, jitter and standard deviation for the latency (right part of Table 20.2). The mean for the one-shot idle state is 4.484 µs and nearly equals the mean of the periodic mode for the same state with 4.742 µs, and the jitter is very low between 0.631 µs and 0.762 µs. Increasing the system load almost doubles the latency for the periodic mode and more than doubles it for the one-shot mode. At the same time, the jitter increases by up to 11.536 µs for the one-shot mode and 7.754 µs for the periodic mode. Increasing the system load also increases the standard deviation SD for both modes. It is remarkable that the standard deviation for the load state in one-shot mode amounts more than 50% of the mean for the idle phase. The corresponding histograms for the values in Table 20.2 are shown in Fig. 20.7 for the idle and load state of the system. The figures are confirming our theory that the

Table 20.2. Quartiles, Jitter and standard deviation for the latency (µs)

RTAI	1.Q	2.Q	3.Q	QD	Min	Max	Mean	Jitter	SD
One-shot idle	4.20	4.30	4.40	0.2	0.6	45.7	4.484	0.6318733	0.7949046
Periodic idle	4.50	4.50	4.60	0.1	0.7	43.7	4.742	0.7626195	0.8732809
One-shot load	8.20	10.20	11.90	3.7	3.7	51.6	10.24	11.53628	3.396510
Periodic load	5.80	6.70	8.20	2.4	1.4	38.3	7.586	7.754853	2.784754

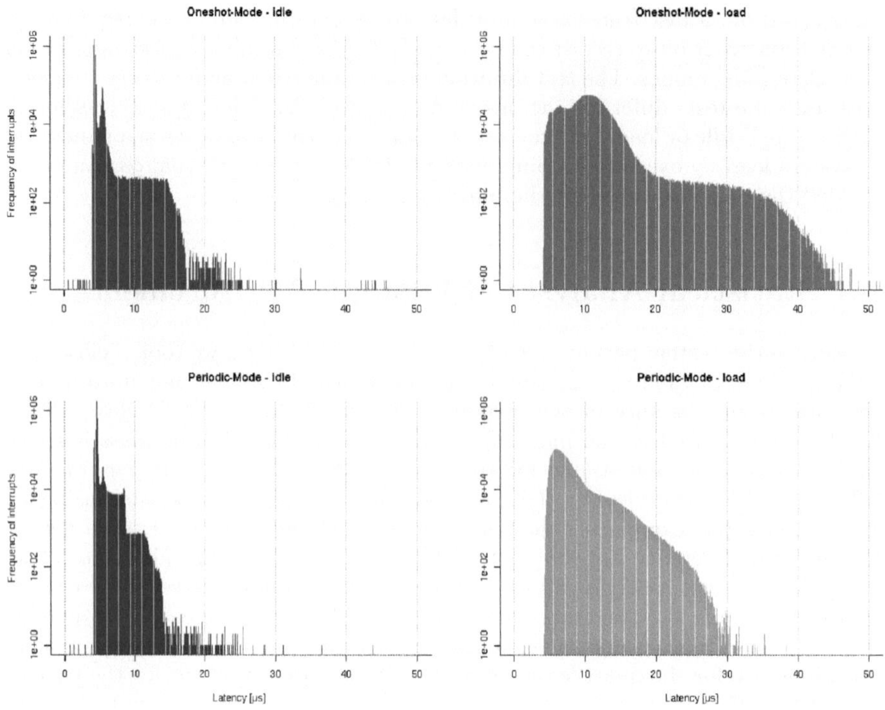

Fig. 20.7. Histogram of the interrupt latency for different system states and scheduling modes

periodic scheduling mode is the better choice, independent of the systems state. The y-axis of the figures is scaled logarithmically and is representing the frequency for the interrupts and the x-axis is representing the systems' response time for an interrupt. Figure 20.7 also repeats the results of Table 20.2 for the marginal difference between the one-shot and periodic mode for the idle state. On the other hand, when the system load increases, the histogram of the distribution for the latency in the periodic mode is slimmer in comparison to the one-shot mode and has obviously a lower mass for its distribution.

20.8 Conclusion

We patched a Linux kernel with a real-time extension to evaluate the performance of an embedded real-time operating system on an industrial PC/104 board. For the performance analyses, we implemented a service routine for the parallel port using the specific real-time API to generate system responses for external caused interrupts. The interrupts are triggered via a signal generator connected to the parallel port of the PC/104 board. We observed the signal curves of the interrupt and the response at the same time on a 4 channel oscilloscope and found out that our embedded system is fast enough for in time response for interrupts up to 100 kHz when the system is in idle state. Increasing the system load also increased the response time and disabled it for interrupts higher 83.5 kHz. We introduced a high-precision counter to gather time-related data for the latency for statistical analyses. The analysis of the latency in different scheduling modes confirmed our theory that the periodic mode in a real-time system gets a better performance nearly every time. Our measurement infrastructure can be easily adapted to evaluate other embedded real-time systems.

References

1. Kemal Köker, Richard Membarth, Reinhard German, Performance analyses of embedded real-time operating systems using high precision counters, Proceedings of The 3rd International Conference on Autonomous Robots and Agents, New Zealand, pp. 485–490,(2006)
2. Hielscher, Kai-Steffen Jens ; German, Reinhard, A Low-Cost Infrastructure for High Precision High Volume Performance Measurements of Web Clusters, In: Peter Kemper ; William H. Sanders: Proc. 13th Conf. on Computer Performance Evaluations, Modelling Techniques and Tools (TOOLS 2003 Urbana, IL, USA September 2-5, 2003), pp. 11–28, 2003
3. Thomas Elste, Untersuchungen zum Aufbau eines echtzeitfhigen Embedded-Linux-Moduls, TU Ilmenau, Institut fr Theoretische und Technische Informatik, Diplomarbeit, 2005
4. PC/104 EMBEDDED CONSORTIUM, PC/104-Specifikation v2.5, http://www.pc104.org, visited on 12/06/2004
5. ARBOR INC, PC/104-module Em104-n513/VL, http://www.arbor.com.tw, visited 12/12/2005
6. LINUX Kernel Archives, http://www.kernel.org/, visited 11/11/2005
7. Erik Andersen, BusyBox: The Swiss Army Knife of Embedded Linux, http://busybox.net, visited 15/07/2005
8. Erik Andersen, uClibc: A C library for embedded Linux, http://www.uclibc.org, visited 15/7/2005
9. Dropbear, Small SSH2 Client, "Home Page", http://matt.ucc.asn.au/dropbear/dropbear.html, visited 15/07/2005
10. RTAI PROJECT, Real-Time Application Interface, http://www.rtai.org, visited 04/10/2005
11. RTAI PROJECT, RTAI API, http://www.rtai.org/documentation/vesuvio/html/api, visited 04/10/2005
12. Various Authors, Adaptive Domain Environment for Operating Systems, http://home.gna.org/adeos/, visited 4/10/2005
13. ELC Frequency generator GF 2006, http://elc.annecy.free.fr, visited 15/12/2005
14. LECROY CORPORATION, LeCroy WaveSurfer 424, http://www.lecroy.com, visited 23/11/2005
15. STRUCK INNOVATIVE SYSTEME GMBH. SIS3820 Multi Purpose Scaler, http://www.struck.de/sis3820.htm, visited 11/12/2005

21

Simple Biped Walking Robot for University Education Considering Fabrication Experiences

Yoshihiko Takahashi, Hirofumi Takagi, Yoshiharu Uchiyama, and Takumi Takashima

Department of System Design Engineering
Kanagawa Institute of Technology, Atsugi
Kanagawa, Japan
ytaka@sd.kanagawa-it.ac.jp

Summary. We have developed a simple six-axis biped robot system to be used as teaching material for undergraduate university students wishing to study an intelligent robot. This system enables students to fabricate the entire robot system by themselves as they understand each individual step and study level from mechanical design to programming languages. The components used to fabricate the robot system are very affordable allowing more students access to the project. This chapter will present the details of the step-by-step study scheme using basic circuit theory, the six-axis biped robot, and lectures.

Keywords: Biped robot, teaching material, university education, step by step study.

21.1 Introduction

Many university students studying engineering courses wish to fabricate an intelligent robot. It is very difficult, however, to fabricate such an intelligent robot for some undergraduate students, who have not yet completed key subjects such as mechanical design, electrical circuit design, control systems, and programming languages. Undergraduate students do not have many development opportunities. It is therefore very important to prepare an opportunity to fabricate an intelligent robot for every student who wishes to develop an intelligent robot.

Many researchers have developed teaching materials of robot fabrication. The line trace robot, for example, was proposed by the Shibaura Institute of Technology [1] and Tokai University [2]. This is a system which enables students to fabricate an entire robot system from mechanism to programme. These systems are very simple and often used as teaching material for children [3]. We believe the line trace robot is suitable for lower level study, but the movement and mechanisms are much too simple for students wishing to study an intelligent robot. Researchers have reported

Y. Takahashi et al.: *Simple Biped Walking Robot for University Education Considering Fabrication Experiences*, Studies in Computational Intelligence (SCI) **76**, 181–188 (2007)
www.springerlink.com

that university students in engineering courses are more interested in studying a humanoid type biped robot [4].

We therefore propose humanoid type biped robots as teaching materials for university students. The University of Tokyo is currently using a very expensive humanoid robot, the HRP2 [5], and The Shibaura Institute of Technology and ZMP Corporation have proposed the e-nuvo [4] which is less expensive than the HRP2. Both of these robot systems have their own control software and hardware systems, and students are able to study intelligent, multi-degree of freedom (DOF) movement; however, they can only use a standard control programme, not one student may design one of their own. Another downside of these models is that they do not allow the student to study mechanical or electrical circuit design.

The purpose of our research is to provide this intelligent robot system as teaching material which will allow students to fabricate the whole robot system from mechanism to software by themselves. We presented a simple six-axis biped robot as a teaching material in which a university student can study robotics with step-by-step fabrication experiences [[6–8]].

This chapter presents a newly developed basic study circuit and an improvement of the six-axis biped robot to ensure and ease the step-by-step study and fabrication. Finally, we introduce the atmosphere of a university lecture in the process [9].

21.2 Target Components of Proposed Simple Six-Axis Biped Robot

This section individually explains the details of the target components of our robot system.

1. The target robot system is a humanoid type biped robot with six motors.
   Many students in engineering courses wish to develop a multi-DOF intelligent robot [4]. A humanoid biped robot is an example of a multi-DOF intelligent robot. A student can perform intelligent motion by using a complete humanoid robot, however the humanoid type biped robot in our research uses only six motors in order to reduce the fabrication time and cost. The intelligent motion in this case implies the computer controlled multi-DOF motion using a multi-DOF robot. Once a student completes the six-axis system, they may proceed to building a 12-axis system in which the robot can turn left or right.
2. Students may fabricate the entire robot system by themselves; the mechanism, the control circuit, the software programme, etc. It is essential to fabricate a whole robot system to understand Robotics. By using the proposed simplified humanoid biped robot, a student conducts the mechanical design using purchased geared toy motors and aluminum plates, and conducts the control circuit design using a PIC microcomputer [[10–12]] and a motor driver. The control software will be programmed using C, the standard programming language for robot controllers.
3. Students perform their fabrication confirming their comprehension level at each step focusing on essential items and key subjects. A student may study and fabricate each item in steps since the fabrication process is divided into small items. A student is required to study many key subjects to fabricate an intelligent robot including mechanical design, electrical circuit, control systems, and

software programming. The study of the key subjects has generally not been completed by underclassmen, therefore, students study only essential items necessary for our robot system. Feedback control is, for example, conducted using the output of a potentiometer in order to understand control systems.

4. Key components are affordable. Among the most important factors for this teaching material is the fabrication cost. When using an expensive robot, students are unable to overhaul the mechanism, or many students have to use one robot system. With affordable parts, students are able to purchase all parts, and fabricate their robot systems individually. This also allows students to overhaul their own robot systems.

21.3 Basic Studies by Basic Study Circuit

The basic study circuit is designed to ensure the step-by-step study scheme. Figure 21.1 illustrates a scheme using the basic study circuit in which a student studies: LED lighting control (step 1), motor ON/OFF control (step 2), A/D input, motor feedback control (step 3), and communication between two PICs (step 4).

Figure 21.2 exhibits the control system of the basic study circuit. A master PIC sends instructions to a slave PIC, which then receives the instruction. The slave PIC inputs switch information, and outputs signals to an LED circuit. The slave PIC controls a motor system by outputting a signal to a motor, and by inputting an analogue signal from a potentiometer. Figure 21.3 demonstrates the fabrication sequence of the basic study circuit. A student fabricates the basic study circuit from step 1 to step 4 by adding necessary parts. In step 2, a motor gearbox and a motor are added. In step 3, two potentiometers are added. In step 4, a master PIC is added.

Figure 21.4 exhibits the communication scheme between the master and slave PICs. Here, very simple parallel three-bit communication is proposed. The table shows the three-bit communication in which slave input values are seven patterns from "000" through "111." The slave PIC controls the motor system depending on the slave input values. For example, the motor angle is controlled to 0°, the center position, when the slave input value is "000." Figure 21.5 displays the experimental results of motor control using the basic study circuit.

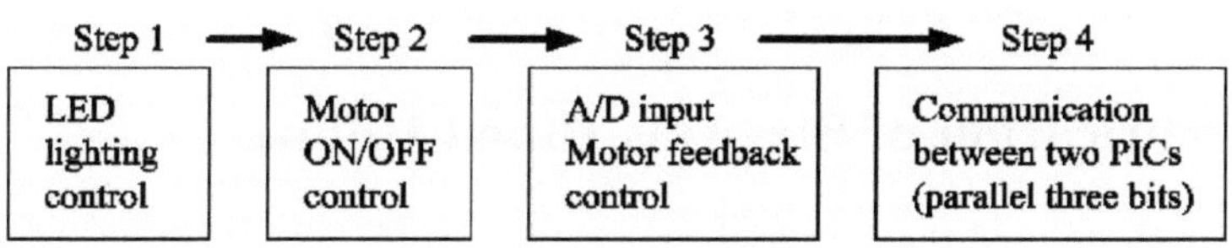

Fig. 21.1. Study scheme of basic study circuit

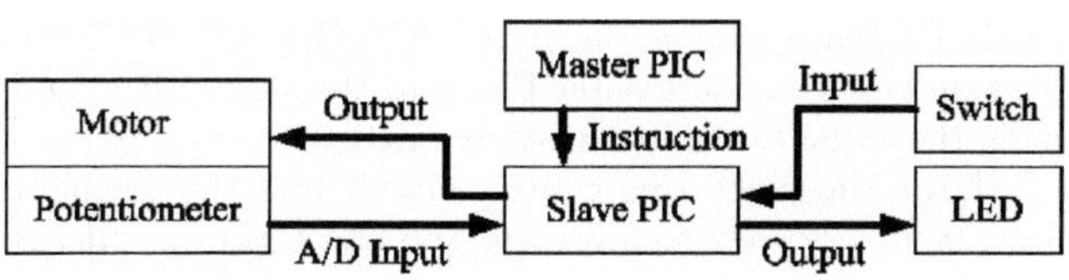

Fig. 21.2. Control system of basic study circuit

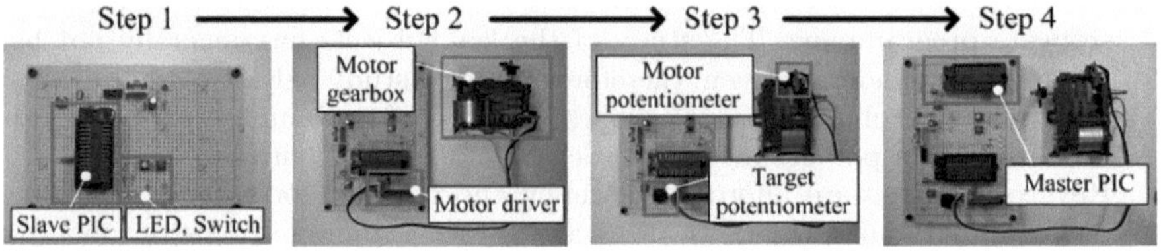

Fig. 21.3. Fabrication sequence of basic study circuit

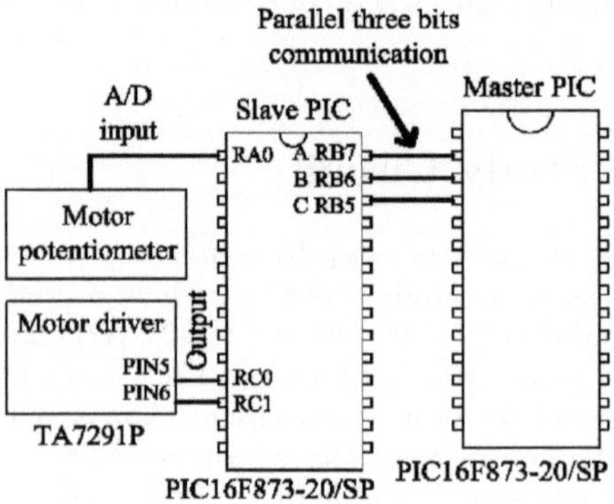

	Slave input A B C	Desired value (max 1023)	Desired voltage [V]	Motor angle [deg]
(0)	0 0 0	500	2.44	0
(1)	0 1 1	350	1.71	-33
(2)	0 1 0	400	1.96	-22
(3)	0 0 1	450	2.20	-11
(4)	0 0 0	500	2.44	0
(5)	1 0 1	550	2.69	11
(6)	1 1 0	600	2.93	22
(7)	1 1 1	650	3.18	33

Fig. 21.4. Communication scheme between the master and slave PICs

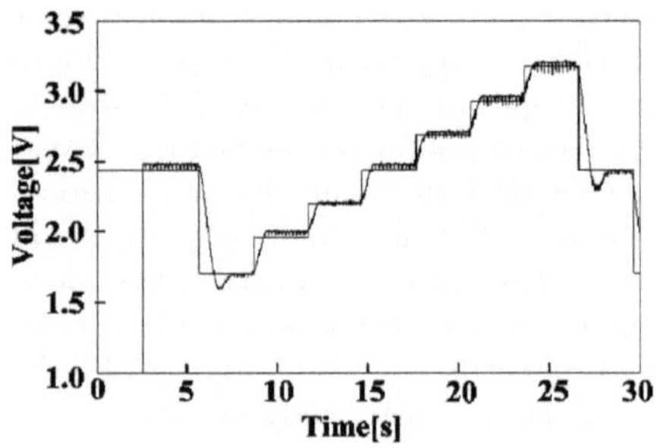

Fig. 21.5. Experimental results of motor control

21.4 Fabrication of Six-Axis Biped Robot

Students may proceed their study to assembly of the six-axis robot after fabricating the basic study circuit. Figure 21.6 shows the control system of the six-axis biped robot. The control system utilizes seven PICs where one PIC is used as the master PIC, and other six PICs are used as the slave PICs. One slave PIC controls one axis. The communication between the master PIC and the slave PICs, and the each axis control scheme is the same as the basic study circuit.

Figure 21.7 shows the fabricated control circuit and the mechanism of the six-axis biped robot. It is very easy to fabricate and check the circuit because all parts are arranged two dimensionally. As affordability is one of the most important factors for the teaching material, affordable components are used in our robot. The total

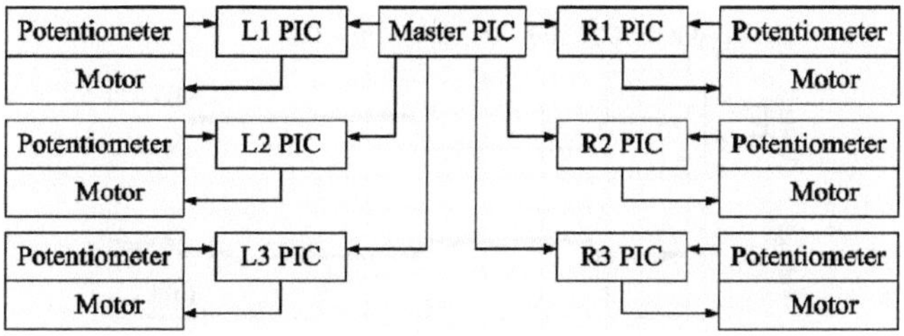

Fig. 21.6. Control system of six-axis biped robot

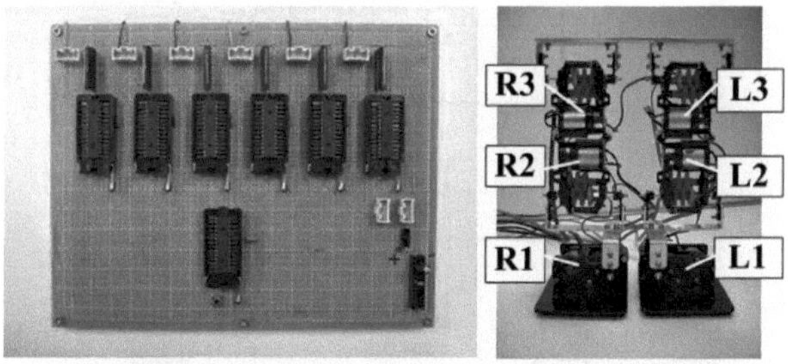

(a) Control circuit (b) Mechanism

Fig. 21.7. Fabricated six-axis biped robot

fabrication cost of one robot system is less than 250 US dollars. Students are able to purchase all of the components by themselves.

Figure 21.8 illustrates the experimental results of walking motion of the six-axis biped robot where in 0 the robot is standing upright, and it conducts one step walking (1 through 7). Figure 21.9 shows the photographs of the walking motion.

21.5 Lecture and Robot Contest Using Six-Axis Biped Robot by Undergraduate University Students

The six-axis biped robot was used and evaluated in the "System Design Project" course in the Department of System Design Engineering at the Kanagawa Institute of Technology. There were 25 students in the course. They were all able to fabricate the robot system, and they all felt that they benefited from the project in their understanding of robot systems and robotics.

A robot contest was also conducted. Figure 21.10 displays photographs from the robot contest. Students were evaluated on walking time along a 20-cm course, and their design report. The robot contest was an excellent goal for the students. Student evaluations confirmed that the robot system is suitable as teaching material.

21.6 Fabrication of Twelve-Axis Robot

Students may proceed their study to assembly of a twelve-axis robot after fabricating the six-axis robot. The control system is using four PICs where one PIC is used as

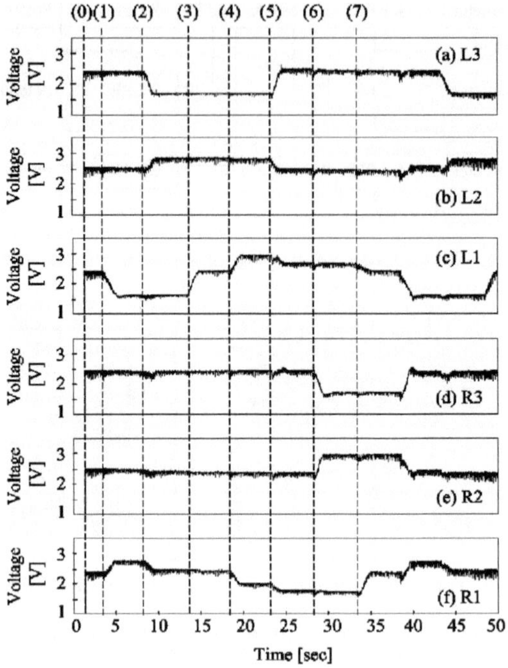

Fig. 21.8. Experimental results of walking motion

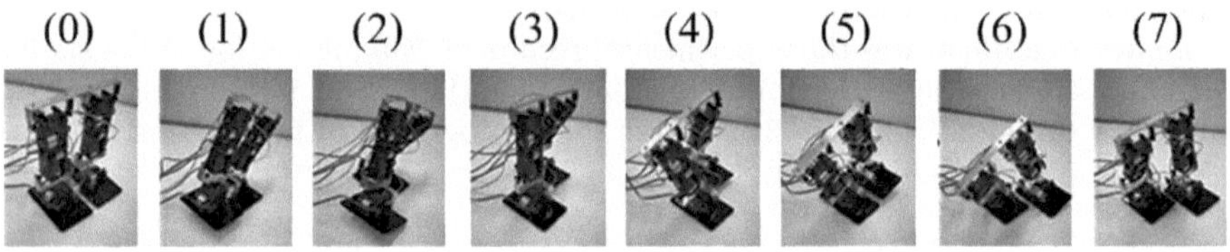

Fig. 21.9. Photographs of walking motion

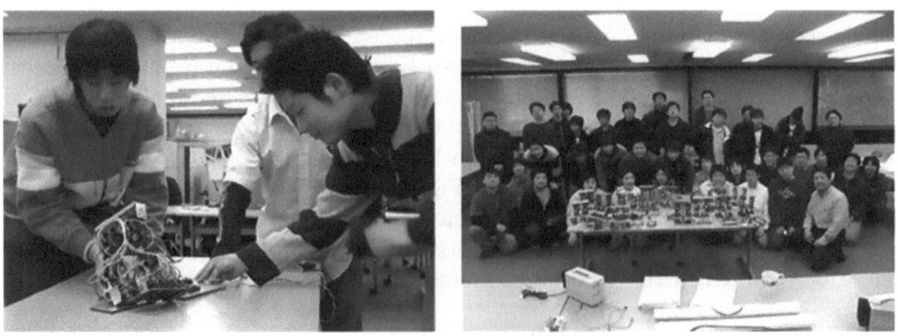

Fig. 21.10. Photographs of robot contest

the master PIC, and other three PICs are used as the slave PICs. One slave PIC controls four axes using parallel programming techniques. An I^2C communication scheme is used to communicate between the master PIC and the slave PICs.

21.7 Conclusions

We have designed a small six-axis biped robot system aimed to be used as teaching materials at universities offering robotic studies. The system can be fabricated using affordable components; geared toy motors, PIC microcomputers, etc. The proposed fabrication scheme will facilitate students' understanding of robotics in simple steps.

The robot system was utilized in a course offered by the Department of System Design Engineering at the Kanagawa Institute of Technology, and in a robot design contest. Students fabricated an entire robot system from mechanism to software by themselves and felt that they benefited their studies with this project. We are preparing further the next step study teaching material of a high grade, twelve-axis robot which can turn left or right.

References

1. Y. Tomimasu, M. Tokizawa, Y. Ando, M. Mizukawa, C. Kasuga, and Y. Ogawa, "Development of the line trace robot as teaching materials for mechatronics education", *Proceedings of 2004 JSME Conference on Robotics and Mechatronics*, Nagoya, Japan, 2P1-L2-36 (2004).
2. N. Komine, K. Inagaki, T. Inaba, and R. Matsuda, "Creative design practice" the subject of robot design for the first year students", *Proceedings of 2004 JSME Conference*, Nagoya, Japan, 2P1-L2-35 (2004).
3. S. Yura, N. Hiraoka, H. Sogo, S. Fujusawa, K. Shikama, S. Urushihara, and H. Tarao, "Kid's KARAKURI workshop at Takamatsu NCT", *Proceedings of 2004 JSME Conference*, Nagoya, Japan, 2P2-L2-36 (2004).
4. O. Yokoyama, M. Mizukawa, and R. Sakai, "e-nuvo: Development of education materials of bipedal locomotion robot", *Proceedings of 2004 JSME Conference*, Nagoya, Japan, 2A1-L2-38 (2004).
5. K. Okada, M. Inaba, and H. Inoue, "Mechano-informatics seminar for undergraduate students using the HRP2 humanoid robot", *Proceedings of 2004 JSME Conference*, Nagoya, Japan, 2P2-L2-39 (2004).
6. Y. Takahashi and M. Kohda, "Simple humanoid biped robot with PIC microcomputer for university education", *Journal of Robotics and Mechatronis*, Vol. 17, No. 2, pp. 226–231 (2005).
7. Y. Takahashi, "Development of biped walking robot kit for university education", *Journal of Robotics Society of Japan*, Vol. 24, No. 1, pp. 16–19 (2006).
8. Y. Takahashi, *Seisaku jiisyu de manabu robotics nyumon*, Ohmsha, Tokyo (2006).
9. Y. Takahashi, H. Takagi, Y. Uchiyama, and T. Takashima, "University Robotics Education with Fabrication Experiences of Simple Biped Robot", *Proceedings of Int. Conf. on Autonomous Robots and Agents, Palmerston North, New Zealand*, pp. 165–170 (2006).

10. T. Endo, *Wakaru PIC micon seigyo*, Seibundo shinkou sya, Tokyo (2002).
11. M. Suzuki, *Tanoshiku dekiru C& PIC seigyo Jikken*, Tokyo denki daigaku syuppankyoku Tokyo (2003).
12. T. Gokan, *C gengo ni yoru PIC programming nyumon*, Gijutsu hyoronsya, Tokyo (2002).

22

Autonomous Stride-Frequency and Step-Length Adjustment for Bipedal Walking Control

Lin Yang[1], Chee-Meng Chew[1], Aun-Neow Poo[1], and Teresa Zielinska[2]

[1] Department of Mechanical Engineering, National University of Singapore
yanglin,mpeccm, mpepooan @nus.edu.sg
[2] Inst. for Aircraft Engineering and Applied Mechanics, Warsaw University of Technology teresaz@meil.pw.edu.pl

Summary. This work focuses on the stride-frequency and step-length autonomous adjustment in response to the environment perturbations. Reinforcement learning is assigned to supervise the stride-frequency. A simple momentum estimation further promised the adjustment. In the learning agent, a sorted action-choose table instructed the learning to find out the proper action in a straightforward way. Incorporating the step-length real-time adjustment mode, the biped is able to smoothly transit motions and walk adaptively to the environment. Dynamic simulation results showed that the supervision is effective.

Keywords: reinforcement learning, bipedal robot, perturbation.

22.1 Introduction

The ultimate tangible goal of the research presented here is to achieve stable bipedal walking adaptive to the environment using recursive search and computations with accessible robot's dynamics.

Bipedal locomotion control is a highly nonlinear and multiaspect control issue. Although some research work has been shown effective for giving a stable walking, walking adaptively to the environment change is still a challenge for further explorations. Among the developed methods, central pattern generator (CPG) concept inspired methods are expected to be flexible with environment. Current popular CPG models such as neural oscillators [1] and Van der Pol equations [2] use nonlinear coupled equations to elaborate the leg harmonious patterns. However, so far CPG models still have not been clearly depicted about their feedback pathways and the relationship between the oscillator models and robot dynamics.

The presented work investigated the issue about bipedal walking gait real-time adjustment under perturbations. It is based on a newly proposed motion generator Genetic Algorithm Optimized Fourier Series Formulation (GAOFSF) [3] [4], which partitions a motion control into a low-level motion generation, and a high-level

L. Yang et al.: *Autonomous Stride-Frequency and Step-Length Adjustment for Bipedal Walking Control*, Studies in Computational Intelligence (SCI) **76**, 189–198 (2007)
www.springerlink.com

motion supervision. The low-level part [3] [4] generated optimized walking solutions and contained parameters for online adjustment [5]. For the high-level motion surveillance, a reinforcement learning (RL) algorithm is used as the core agent to supervise the walking pace to be adaptive to perturbations. Therefore, with those two levels working together, the GAOFSF model [4] gives the behavior of a pattern generator but also has clearer feedback pathways.

In this chapter, we focus on the high-level supervision for walking under perturbations. The target robot is modeled as the same as the one modeled in a previous work report [3].

22.2 Stride-Frequency Online Adjustment

This section discusses the automatic stride-frequency adjustment using the Truncated Fourier Series (TFS) model. Stride-frequency is defined as the fundamental frequency ω_h in the TFS model [3]. Since walking speed has been optimized even using the GAOFSF, this stride-frequency can represent the average walking speed. The main use of the stride-frequency adjustment is to maintain the kinetic energy under perturbations. The accessible range of the adjustment can be referred to our previous chapter [5].

22.2.1 Modules Included in the Reinforcement Learning (RL)

The following paragraphs will present the modules considered in the supervisory controller for the stride-frequency adjustment.

Learning Agent

Reinforcement learning (RL) is a class of learning problem in which an agent learns to achieve a goal through trial-and-error interactions with the environment. The learning agent learns only from reward information and it is not told how to achieve the task. From failure experience, the learning agent reinforces its knowledge so that success can be attained in future trials. The environment is modeled to be a fully observable Markov decision process (MDP) that has finite state and action sets. Q-learning method recursively estimates a scalar function called optimal Q-factors ($Q^*(i,u)$) from experience obtained at every stage, where i and u denote the state and corresponding action, respectively. The experience is in the form of immediate reward sequence, $r(i_t, u_t, i_{t+1})(t = 0, 1, 2\cdots)$. ($Q^*(i,u)$) gives the expected return when the agent takes the action u in the state i and adopts an optimal policy π^* thereafter. Based on ($Q^*(i,u)$), an optimal policy π^* can easily be derived by simply taking any action u that maximizes ($Q^*(i,u)$) over the action set $U(i)$. Equation (22.1) gives the single-step sample update equation for $Q(i,u)$ [6]:

$$\begin{aligned} Q_{t+1} \leftarrow & Q_t(i_t, u_t) + \alpha_t(i_t, u_t)[r(i_t, u_t, i_{t+1}) \\ & + \gamma max_{u \in U(i_{t+1})} Q_t(i_{t+1}, u) - Q_t(i_t, u_t)] \end{aligned} \tag{22.1}$$

where the subscript indicates the stage number; $r(i_t, u_t, i_{t+1})$ denotes the immediate reward received due to the action u_t taken which causes the transition from state i_t to i_{t+1}; $\alpha \in [0,1)$ denotes the step-size parameter for the update; $\gamma \in [0,1)$ denotes the discount rate. (1) updates $Q(i_t, u_t)$ based on the immediate reward $r(i_t, u_t, i_{t+1})$ and the maximum value $Q(i_{t+1}, u)$ of over all $u \in U(t_{i+1})$. The convergence theorem for the Q-learning algorithm that uses lookup table representation for the Q-factors is:

Theorem 1. *Convergence of Q-learning: for a stationary Markov decision process with finite action and state spaces, and bounded rewards $r(i_t, u_t, i_{t+1})$. If the update (1) is used and $\alpha_t \in [0,1)$ satisfies the following criteria: $\sum_{i=1}^{\infty} \alpha_t(i,u) = \infty$ and $\sum_{i=1}^{\infty} [\alpha_t(i,u)]^2 < \infty, \forall (i,u)$; then $Q_t(i,u) \rightarrow Q^*(i,u)$ as $t \rightarrow \infty$ with probability 1, $\forall (i,u)$.*

In this chapter, Cerebellar Model Articulation Controller (CMAC) is used as a multivariable function approximator for the Q-factors in the Q-learning algorithm. CMAC has the advantage of having not only local generalization, but also being low in computation. The Q-learning algorithm using CMAC as the Q-factor function approximator is summarized as:

> ***Initialize*** weights of CMAC
> ***Repeat*** (for each trial):
>   ***Initialize*** i
>   ***Repeat*** (for each step in the trial)
>   Select action u under state i using policy (say ϵ) based on $\hat{Q}$.
>   Take action u,
>   Detect new state i' and reward r
>   ***if*** i' not a failure state
> $\delta \longleftarrow r + \gamma max'_u \hat{Q}(i', u') - \hat{Q}(i, u)$
>   Update weights of CMAC based on δ
> $i \longleftarrow i'$
>   ***else*** $(r = r_f)$
> $\delta \longleftarrow r_f - Q(\hat{i}, u)$
>   Update weights of CMAC based on δ
>   ***Until*** failure encountered or target achieved
> ***Until*** target achieved or number of trials exceed a preset limit

State Variables

The input state representing and describing the motion dynamics for the learning agent is composed of: (1) trunk's CG error, (2) external force f, (3) current stride-frequency ω_t; the output state is the normalized change of the stride-frequency $\Delta\omega_h$ with respect to different step-length scales. Such state variables contain the current average walking velocity from the current stride-frequency for a fixed step-length walking (supposing the change of stride-frequency is proportional to its step-length level), motion acceleration which depicts the dynamic trend through external force and the actual stance posture obtained from the trunk's CG error.

Reward Functions

Reward functions give rewards to motions which satisfied the motion objectives. However, it also punishes a failed action. The reward function is shown as (22.2):

$$reward = \begin{cases} k \cdot (C - |\Delta\omega_h|) & \text{for actions succeeded} \\ -k_e|E| - k_t|Tq| & \text{for actions failed} \end{cases} \tag{22.2}$$

k,k_e,k_t are the scales for the performances. C is a positive constant which stands for a fixed reward; E and Tq are the error of trunk's CG and the excessive ankle joint torque. For actions succeeded, there is still a difference considered among different actions. If the change of stride-frequency is small, it means the updated frequency is not deviated much from the original frequency, the action will be given more rewards. This is to encourage the walking tolerance provided the selected action is successful.

Learning Tasks

In the RL supervision, the more experiences the learning agent has gone through, the more reliability the controller can ensure. The advantage is as long as the learning agent is established, it can always keep learning different tasks assigned and increase the experience unboundedly. However, it is not easy to learn enough experiences in a short time. To reduce the chance that the robot might encounter situations which have not been trained yet, 6 representative tasks were assigned to the RL. In this chapter, flat-terrain walking [3] is taken as an example.

First, the stride-frequency ω_h and external force f are identified into groups $\omega_s, \omega_m, \omega_b$ and f_s, f_m, f_b. The subscript s, m, b stands for small, medium, big, respectively. With the learning carried on, the following groups can be further separated into subgroups.

$$\omega_s : \omega_h \in [2.9, 4.2] \longleftrightarrow \omega_t = 3.5 \quad \omega_m : \omega_h \in [4.3, 5.6] \longleftrightarrow \omega_t = 4.9$$

$$\omega_b : \omega_h \in [5.7, 7.0] \longleftrightarrow \omega_t = 6.3 \quad f_s : f \in [10, 30) \longleftrightarrow f_t = 20$$

$$f_m : f \in [30, 50) \longleftrightarrow f_t = 40 \quad f_b : f \in [50, 70) \longleftrightarrow f_t = 60$$

The representative 6 tasks are described as:

1. $\omega_0 = \omega_s \cap f = random(f_s)$ with the objective: continuous walking at least for 4.0 seconds with the frequency increased to ω_b
2. $\omega_0 = \omega_s \cap f = random(f_m)$ with the objective: continuous walking at least for 3.0 seconds with the frequency increased to ω_b
3. $\omega_0 = \omega_s \cap f = random(f_b)$ with the objective: continuous walking at least for 1.2 seconds with the frequency increased to ω_b
4. $\omega_0 = \omega_b \cap f = -random(f_s)$ with the objective: continuous walking at least for 1.5 seconds with the frequency decreased to ω_s
5. $\omega_0 = \omega_b \cap f = -random(f_m)$ with the objective: continuous walking at least for 1.0 seconds with the frequency decreased to ω_s
6. $\omega_0 = \omega_b \cap f = -random(f_b)$ with the objective: continuous walking at least for 0.7 seconds with the frequency decreased to ω_s

Table 22.1. RL action instruction table

State[1]	State[2]	State[3]	Action u
+	+	$\omega_h(min)$	$\uparrow$
+	+	$\omega_h(max)$	$=\uparrow$
+	+	ω_h	$\uparrow$
−	+	$\omega_h(min)$	$\uparrow$
−	+	$\omega_h(max)$	$=\downarrow$
−	+	ω_h	$\uparrow or \downarrow$
+	−	$\omega_h(min)$	$=\uparrow$
+	−	$\omega_h(max)$	$\downarrow$
+	−	ω_h	$\uparrow or \downarrow$
−	−	$\omega_h(min)$	$=\downarrow$
−	−	$\omega_h(max)$	$\downarrow$
−	−	ω_h	$=\downarrow$

Common constraints are: the absolute value of trunk's CG error < 0.05 m (trunk CG height is 0.925 m); the sum of all joint torques in one discrete time step (0.01 s) is less than 10 000 Nm (one discrete-step has 100–200 time refreshes); and the stance foot has no rotation bigger than 10°. Therefore, with the torque constraint getting strict, the energy consumption will be expected reduced; with the extension of the walking time, the tolerance will be improved.

Action-Choose Table

To reduce the number of iterations and choose the correct action easier, a logic instruction is established as shown in Table 22.1:

Momentum Estimation

This is to enhance the confidence for the stride-frequency online adjustment in case the current situation has not been experienced in the CMAC network. Therefore, it is only used when the current state has not been trained by the CMAC network yet. The stability issue and the allowed sudden change of the stride-frequency have been discussed in the previous work [5]. Therefore, as long as the estimated change is within the allowed range, the dynamic stability of this estimation is promised. According to the sensed or calculated external force and the stride-frequency update time-step, the input momentum given by the perturbation can be estimated by (22.3)–(22.5).

$$M = f_t \cdot \Delta t = \Sigma_{i=1}^{7} m_i \cdot \Delta v_i \tag{22.3}$$

$$\Delta v_1 = 0; \Delta v_2 = l_2/L \cdot \Delta v; \Delta v_3 = l_1/L \cdot \Delta v \tag{22.4}$$

$$\Delta v_4 = \Delta v_5 = \Delta v_6 = \Delta v_7 = \Delta v \tag{22.5}$$

M is the input momentum, l_1, l_2, and L are the link-length with effect of upper, lower and the whole leg, respectively. i is the link number starting from the stance foot to the swing foot. The change of stride-frequency is calculated using (22.8):

$$\Delta\omega_h = \frac{\pi \cdot \Delta v}{S} \tag{22.6}$$

S is the step-length and Δv can be calculated through (22.3)–(22.5). Equation (22.6) actually updates the walking to be out of the effect of the external force if the adjust is within the stable range.

22.3 Step-Length Online Update

The presented step-length adjustment is considered externally, not included in the learning control. It is to adjust the position error. Step-length updates itself through the scaling parameter R in the (1) and (2) whenever the investigated phase angle has an error to the phase defined in the transition module [5]. Actually changing the step-length is to maintain the walking pattern according to the stance posture while tuning the stride-frequency is to maintain the kinetic energy. Therefore, with the step-length and stride-frequency online adjustment, robot dynamics becomes more controllable under perturbation or transitions. The basic rule for adjusting the step-length follows the linear interpolation [5]. The further step-length adjustment mode will be presented in the future work.

22.4 Experiments on Walking Under Perturbations

This part presented the learning result and dynamic simulation result of the supervisory control applied to the bipedal walking under perturbations. For the recovery of a pattern, as soon as there is no external force sensed, the robot will be assigned the recovery module which follows the linear approaching.

22.4.1 Reinforcement Learning Results

Figures 22.1 and 22.2 show the Q-learning performances for the six tasks mentioned in Sect. 22.2.1. Since the basic walking pattern has already been regulated, the RL learning task is relatively simple and not time consuming yet.

22.4.2 Dynamic Simulation Experiment

The aforementioned algorithm has been implemented dynamically on a walking experiment described as: external forces $20N$, $40N$, $60N$ were applied from $t = 2s$ to $t = 5s$, $t = 6s$ to $t = 6.8s$, and $t = 9s$ to $t = 9.25s$, respectively; external forces $-20N$,$-40N$,$-60N$ were applied from $t = 11.2s$ to $t = 11.6s$, $t = 13.7s$ to $t = 14.1s$, and $t = 16s$ to $t = 16.3s$, respectively.

With the presented RL agent and the momentum estimation. The robot is able to overcome all batches of perturbation and behaves similar to human beings. Figure 22.3 shows that robot walking was disrupted without any motion adjustment but Figs. 22.4 and 22.5 show the successful walking and the real-time stride-frequency and step-length adjustment information under all perturbations.

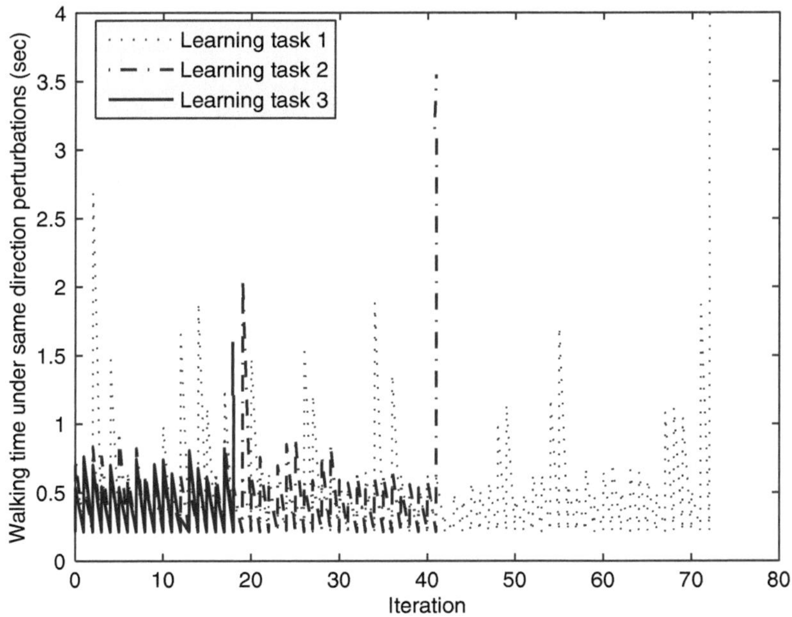

Fig. 22.1. Learning performance of task 1–3

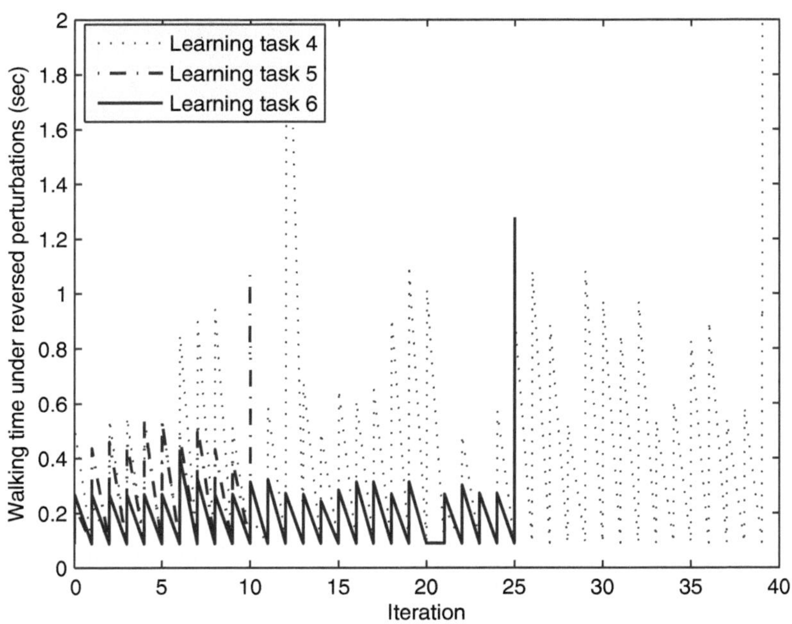

Fig. 22.2. Learning performance of task 1–3

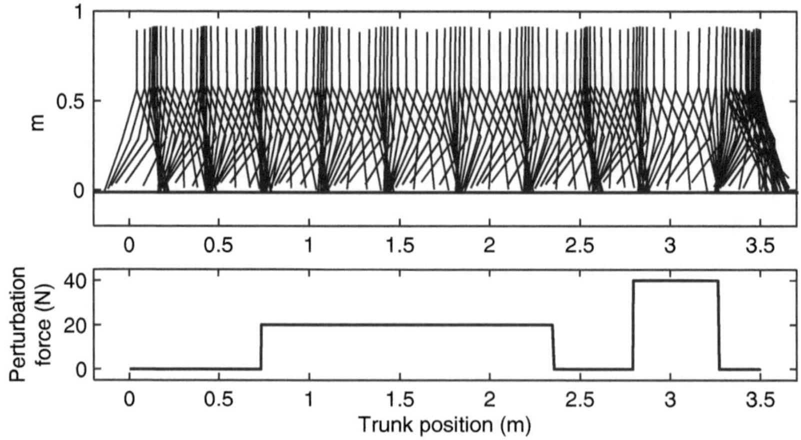

Fig. 22.3. Stick diagram of walking without stride-frequency or step-length adjustment (walking was disrupted finally)

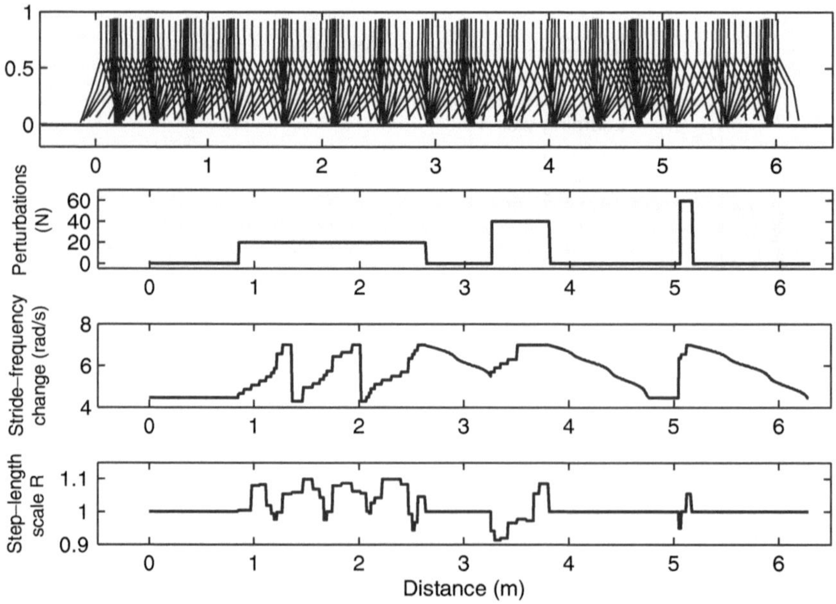

Fig. 22.4. Stick diagram of walking with online stride-frequency and step-length adjustment (under 1st, 2nd, and 3rd batch of perturbations)

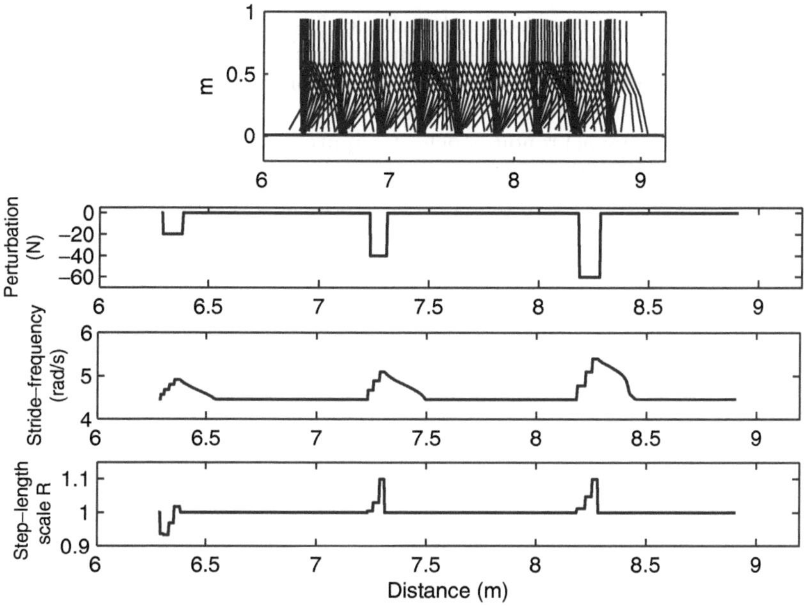

Fig. 22.5. Stick diagram of walking with online stride-frequency adjustment(under the 4th, 5th, and 6th perturbations)

22.5 Conclusion and Future Works

Concluded from the conducted walking experiment, this supervisory control worked well and instructed the robot to get over of all batches of perturbations. As motion pattern has been regulated and optimized at the low-level first using GAOFSF [3] [4], the learning agent did not take a long time to train the RL network. Further discussions regarding less joint torque, trunk position error, and better robustness of this supervisory controller will be presented in the following report.

References

1. Fukuoka Y, Kimura H and Cohen A-H (2003) Adaptive Dynamic Walking of a Quadruped Robot on Irregular Terrain Based on Biological Concepts The Int. Journal of Robotics Research 22(3-4): 187–202
2. Dutra S, Filho P and Romano F (2004) Modeling of a bipedal locomotor using coupled nonlinear oscillators of Van der Pol Bio. Cybernetics 88(4): 286–292
3. Yang L, Chew C-M, Poo A-N and Zielinska T (2006) Adjustable Bipedal Gait Generation Using Genetic Algorithm Optimized Fourier Series Formulation In: IEEE/RSJ Int. Conf. on Intelligent Robots and Systems 4435–4440
4. Yang L, Chew C-M, Zielinska T and Poo A-N (2007) A Uniform Biped Gait Generator with Off-line Optimization and On-line Adjustable Parameters Int. Journal of Robotica accepted

5. Yang L, Chew C-M, Poo A-N and Zielinska T (2006) Adjustable Bipedal Gait Generation using GA Optimized Fourier Series Formulation: Real-time Gait Adjustment In: Proceedings of the Int. Conf. on the Autonomous Robots and Agents (ICARA) 111–116 New Zealand
6. John N-T (1994) Asynchronous stochastic approximation and q-learning Machine Learning 16(3): 185–202

23

Application of Extended Kalman Filter Towards UAV Identification

Abhijit G. Kallapur, Shaaban S. Ali, and Sreenatha G. Anavatti

University of New South
Wales, Australian Defence Force Academy, Canberra
ACT - 2601, Australia
{a.kallapur,s.salman,agsrenat}@adfa.edu.au

Summary. This chapter considers the application of Extended Kalman Filtering (EKF) towards Unmanned Aerial Vehicle (UAV) Identification. A 3 Degree-Of-Freedom (DOF), coupled, attitude dynamics is considered for formulating the mathematical model towards numerical simulation as well as Hardware-In-Loop (HIL) tests. Results are compared with those obtained from a state-space estimation method known as the Error Mapping Identification (EMId) in the context of processing power and computational load. Furthermore the statistical health of the EKF during the estimation process is analysed in terms of covariance propagation and rejection of anomalous sensor data.

Keywords: Extended Kalman Filter, UAV, identification, nonlinear.

23.1 Introduction

Unmanned Aerial Vehicle (UAVs) fly at very low speeds and Reynolds numbers, have nonlinear coupling and tend to exhibit time varying characteristics. In order to control such vehicles it becomes necessary to design and develop robust and adaptive controllers and hence, identify the system. The issues of identification relating non-linear multi-input-multi-output systems are a challenging area of research. State and parameter estimation, an integral part of system identification, has been carried out on various flight data and simulations as evident from literature [1–3]. Though methods such as Least Squares [1], Gauss-Newton [3] and Kalman filters [4] have been used successfully for flight parameter estimation, their performance is known to deteriorate with an increase in system non-linearities. Constant gain filter methods have been used to estimate system coefficients showing satisfactory results for moderately non-linear systems [5]. However, time varying filter methods become essential for systems with higher degrees of non-linearity [3]. The use of Extended Kalman Filter (EKF) in such cases has proved to be a good technique for identification [3, 4]. Here we present results as obtained from numerical simulations and Hardware-In-the-Loop (HIL) towards identifying UAV attitude dynamics using the method of EKF. These results are further compared against results from a more

A.G. Kallapur et al.: *Application of Extended Kalman Filter Towards UAV Identification*, Studies in Computational Intelligence (SCI) **76**, 199–207 (2007)
www.springerlink.com

recent, Error Mapping Identification (EMId) technique. In addition, statistical investigation into the health of EKF in terms of covariance analysis and Chi-Squared analysis are discussed.

The chapter is organised into nine sections. Section 23.2 describes the UAV Attitude Dynamics. Sections 23.3 and 23.4 introduce the concepts of EKF and EMId, respectively. Results from numerical simulations and HIL are discussed in Sect. 23.5. Section 23.6 discusses statistical, health monitoring issues concerning EKF with concluding remarks outlined in Sect. 23.7.

23.2 UAV Attitude Dynamics

The flight data used for identification was acquired from a 3-DOF, Inertial Measurement Unit (IMU) designed and built at UNSW@ADFA. These data are highly noisy as can be seen from Figs. 23.1 and 23.2. In order to cope with various noisy and non-linear characteristics of the UAV, the attitude dynamics considered in this chapter are mapped by a set of three highly coupled, non-linear differential equations [6, 7]:

$$\begin{aligned}
\dot{p} &= \frac{\{I_z\left[L + (I_y - I_z)qr\right] + I_{xz}\left[N + (I_x - I_y + I_z)pq - I_{xz}qr\right]\}}{(I_x I_z - I_{xz}^2)} \\
\dot{q} &= \frac{\left[M + (I_z - I_x)pr + (r^2 - p^2)I_{xz}\right]}{I_y} \\
\dot{r} &= \frac{\{I_x\left[N + (I_x - I_y)pq\right] + I_{xz}\left[L + (I_y - I_x - I_z)pq + I_{xz}pq\right]\}}{(I_x I_z - I_{xz}^2)}
\end{aligned} \tag{23.1}$$

where,

p, q, r : Angular rates of roll, pitch and yaw,
L, M, N : Aerodynamic moments of roll, pitch and yaw,
I_x, I_y, I_z, I_{xz} : Moments of inertia.

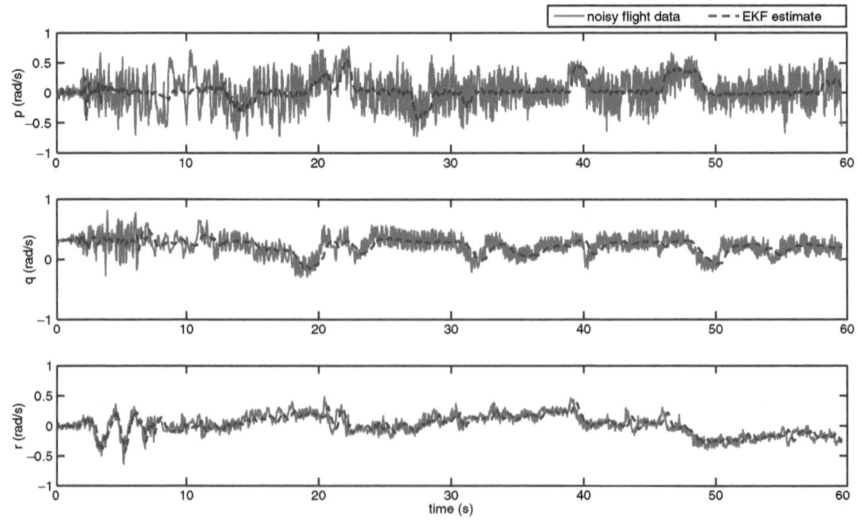

Fig. 23.1. EKF state estimation – numerical simulation

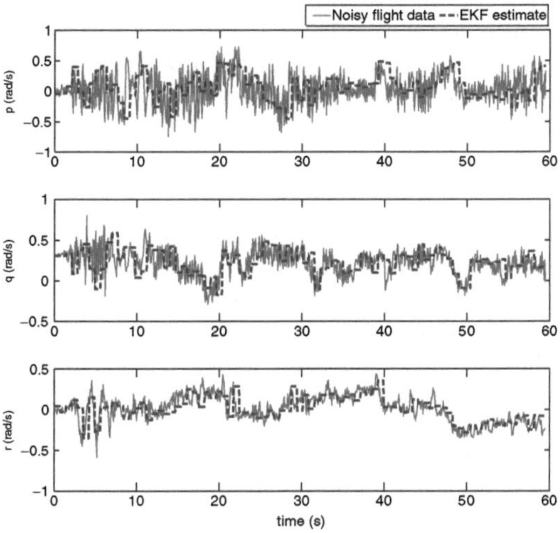

Fig. 23.2. EKF state estimation – HIL

The set of equations described by (23.1) can be represented in the statespace form as:

$$\dot{x}(t) = A(t)f(x, u, t) \tag{23.2}$$

where the state vector $x(t)$, the input vector $u(t)$ and the coefficient matrix $A(t)$ are given by:

$$x(t) = [p\, q\, r]^T \tag{23.3}$$

$$u(t) = [\delta_e\, \delta_r\, \delta_a\, \delta_{th}]^T \tag{23.4}$$

$$A(t) = \begin{bmatrix} A_{1,1} & A_{1,2} & 0 & 0 & 0 & A_{1,6} & 0 & A_{1,8} & 0 & A_{1,10} & A_{1,11} & A_{1,12} \\ 0 & 0 & A_{2,3} & A_{2,4} & A_{2,5} & 0 & A_{2,7} & 0 & A_{2,9} & 0 & 0 & A_{2,12} \\ A_{3,1} & A_{3,2} & 0 & 0 & 0 & A_{3,6} & 0 & A_{3,8} & 0 & A_{3,10} & A_{3,11} & A_{3,12} \end{bmatrix} \tag{23.5}$$

$$f(x, u, t) = \left[pq\ qr\ pr\ p^2\ r^2\ p\ q\ r\delta_e\ \delta_r\ \delta_a\ \delta_{th}\right]^T \tag{23.6}$$

The elements of the matrix $A(t)$ are unknown and need to be identified. A detailed discussion of the elements of the matrix $A(t)$ can be found in [7].

23.3 Extended Kalman Filter (EKF)

The EKF used in this chapter is based on the non-linear state-space process and observation models described by:

$$\left.\begin{aligned} x_{k+1} &= f(x_k, u_k, w_k) \\ z_k &= h(x_k, v_k) \end{aligned}\right\} \tag{23.7}$$

where u_k, w_k and v_k denote the control input, process noise and measurement noise, respectively, at time step k. The state estimation problem can be further extended to state and parameter estimation by including the unknown parameters with the states to construct an augmented state vector [8, 9]. The combined state and parameter estimation can then be modelled as a single-pass procedure using the EKF [3]. The augmented state vector used in this application constitutes of three states and 13 parameters given by [7]:

$$x_a = [p\ q\ r\ L_p\ L_r\ L_{\delta_r}\ L_{\delta_a}\ N_p\ N_r\ N_{\delta_r}\ N_{\delta_a}\ M_q\ M_{\delta_e}\ M_{\delta_t h}\ L_{\delta_t h}\ N_{\delta_t h}\ ] \tag{23.8}$$

The 13 parameters in the above equation provide definitive, physical significance in terms of aerodynamic derivatives and can be used directly to study the aerodynamic stability of the flight data for a certain UAV.

Initial estimates of various states, parameters and covariance are necessary for EKF estimation. These are set to nominal values based on engineering judgement. The diagonal elements of the initial covariance matrix P are set to reflect the confidence in corresponding estimates [3]. For the three states, the covariance values are set to the corresponding measurement error variances [8], while those for the augmented parameters are set to a relatively high value [3].

The process of estimation using EKF consists of two iterative steps: propagation and correction. The linear, discrete model of equation 23.7 is used to propagate states and covariance to the next time-step based on the current knowledge of the model, given by [2]:

$$\left.\begin{aligned} \hat{x}_k^- &= f(\hat{x}_{k-1}^+, u_{k-1}, t_{k-1}) \\ P_k^- &= F\, P_k^+ F^T + Q_k \end{aligned}\right\} \tag{23.9}$$

Continuous propagation leads to accumulation of model-induced errors and hence the need for a correction step. Corrections are generally made as and when new measurements are available [2, 7].

$$\left.\begin{aligned} K_k &= P_k^- H^T (H\, P_k^-\, H^T + R_k)^{-1} \\ \hat{x}_k^+ &= \hat{x}_k^- + K_k (z_k - H\, \hat{x}_k^-) \\ P_k^+ &= (I - K_k\, H) P_k^- \end{aligned}\right\} \tag{23.10}$$

Though EKF simultaneously provides state as well as parameter estimation for non-linear, time-variant and unstable systems [2, 7], it comes at its own cost. The filter performance and divergence is highly sensitive to initial estimates. Moreover, output precision can be severely hampered in case of highly non-linear systems due to the first order linearisation. As can be seen from various equations mentioned in this section the computational requirements of EKF can be quite demanding. This can be controlled to a certain extent by reducing the frequency of correction, but as a trade-off for estimation efficiency.

23.4 Error Mapping Identification (EMId)

The non-linear EMId method is a robust, direct estimation technique based on state-space identification [6, 7, 10]. The system under consideration is of the form:

$$\dot{x}(t) = F(x, u, t),\ t \geq 0\ x(t_0) = x_0 \tag{23.11}$$

which can be written in the matrix state space form as [6, 10]:

$$\dot{x}(t) = A(t)\, f(x, u, t),\ t \geq 0\ x(t_0) = x_0 \tag{23.12}$$

where $A(t) \in R^{cxn}$ is the real matrix and $f(.)$ denotes a given real analytic function.

At a given flight condition, UAV dynamics can be assumed to be time invariant, i.e. the system parameters are constant. Hence the system in (23.12) becomes time invariant, so $\dot{A}(t) = 0$. This leads to the non-linear equation:

$$\dot{A}_m(t) = [\Delta\dot{x}(t) - A_m\ \Delta f(x, x_m, u, t)]\ f(x, u, t)^T\ K\ \ A_m(t_0) = A_{m0} \tag{23.13}$$

where Δ denotes the error between actual and estimated values, subscript m denotes identified values and K represents the weighting matrix. The unknown parameters are found by solving the non-linear differential equation (23.13) [6, 7].

23.5 Results

Numerical simulations are a good platform to check for the operation of the algorithm, which in our case were executed using ***MATLAB®Simulink®. The HIL simulations on the other hand provide near-realistic implementation results where the actual UAV is replaced by a computer simulation, while the autopilot is used as it would have been onboard the UAV. Owing to restrictions imposed by the processing power of the PC-104 unit the sampling rate as well as correction frequency for the EKF was reduced in case of HIL.

Figures 23.1 and 23.2 provide results for p, q, r estimation using EKF as obtained from numerical simulations and HIL. It is evident from them that the EKF estimation averages noise effects efficiently due to its inherent filtering properties. Detailed results for the EMId process can be found in [7]. Figure 23.3 on the other hand compares coefficient estimates of EKF and EMId.

Table 23.1 summarises the root mean square error (RMSE) values obtained from the state estimations for both numerical simulations as well as HIL. It is clear from this table that the EKF estimates were better than EMId estimates for numerical simulations, whereas EMId performed better than EKF in case of HIL simulations (apart from q estimates). This variation in performance can be attributed to the fact that EKF was operated at a higher sampling rate and the states and parameters were updated every single time step in case of the numerical simulation. However, in case of the HIL they were updated once every ten time steps and the filter (EKF) was operated at a lower sampling rate. This was seen as a processor related restriction imposed due to high computational requirement of the EKF. EMId on the other hand had no change in its computational method for both numerical as well as HIL simulations.

23.6 EKF Health Monitoring

Monitoring filter health and rejecting anomalous sensor inputs are critical to the operation of Kalman filters. The formulation of Kalman Filter encapsulates equations that aid in various statistical analyses. Here we discuss statistical methods

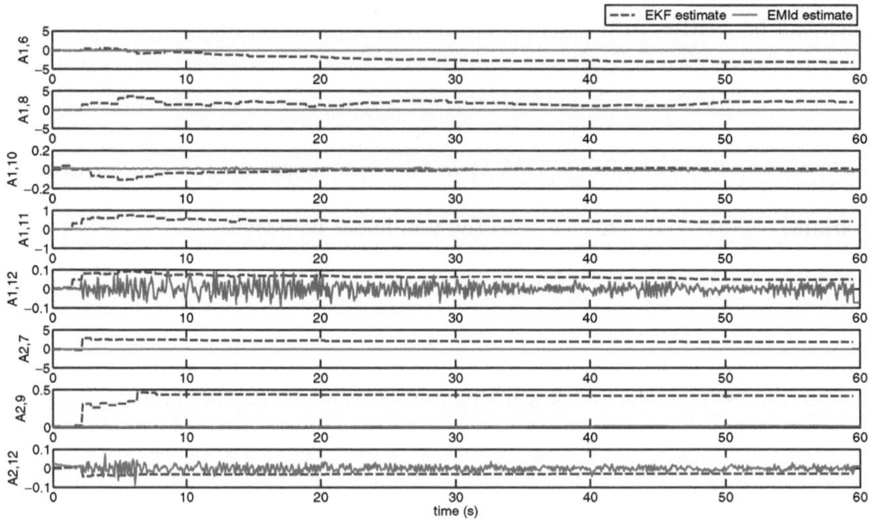

Fig. 23.3. EKF vs. EMId – co-efficient comparison

Table 23.1. RMSE comparison of state estimation

Process	Method	p_{RMSE}	q_{RMSE}	r_{RMSE}
Numerical simulation	EMId	0.2120	0.1445	0.0741
	EKF	0.1580	0.0477	0.0311
HIL	EMId	0.2837	0.1758	0.0864
	EKF	0.2925	0.1565	0.1129

that aid in the process of detecting anomalous sensor data and keeping track of the growth in estimation uncertainties for the three main states p, q and r.

23.6.1 Rejecting Anomalous Sensor Data

Though Kalman Filters have the inherent property of noise rejection they cannot deal with anomalous sensor data that are introduced as a result of signal corruption and sensor failure [11]. It is important to detect such errors and eliminate them before letting them corrupt the estimation process. The Kalman Gain equation K provides the information matrix of innovations as in (23.10), which along with the measurement residual can be used to detect anomalous sensor data that lie within a certain threshold for operational stability.

$$Y_k = (H\, P_k^-\, H^T + R_k)^{-1} \tag{23.14}$$

$$\xi_k = z_k - H_k\, \hat{x}_k^- \tag{23.15}$$

The statistical inference can be obtained from Chi-Squared analysis which makes use of (23.14) and (23.15), and is given by [11]:

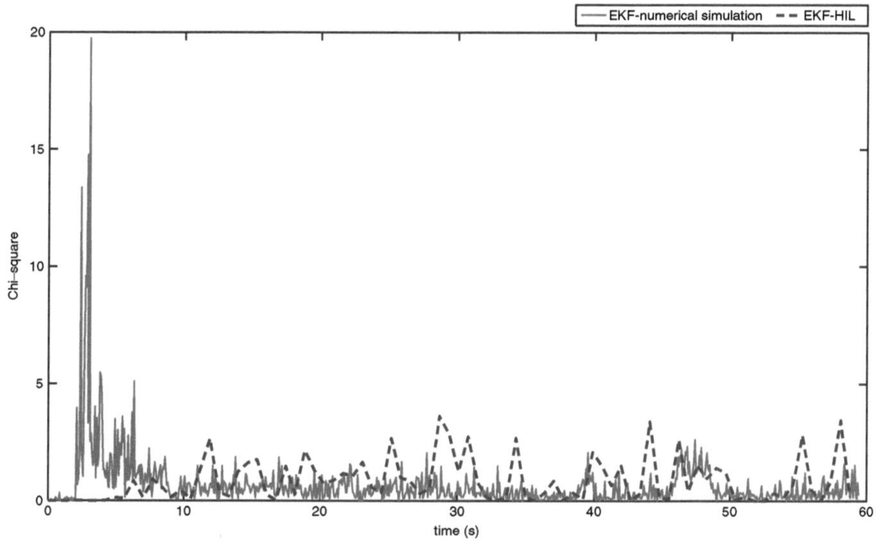

Fig. 23.4. EKF – chi-square distribution

$$\chi^2 = \frac{\xi_k \, Y_k \, \xi_k^T}{l} \tag{23.16}$$

where l specifies the number of measured states. Figure 23.4 shows a plot of Chi-Squares as obtained for the EKF estimation process. Using the Chi-Squared distribution table the probability of a certain chi-square distribution to lie within a specified threshold of operational stability can be found. In our case 70.93% of data for the HIL and 87.42% of data for the numerical simulations were found to exist within a stability threshold of 75%.

23.6.2 Covariance Analysis

Covariance analysis is an important test to keep track of the growth in uncertainties in the estimation procedure. This is achieved by following the change in the diagonal elements of the covariance matrix P. For simplicity the results from covariance analysis for the three main states p, q and r, are depicted in Fig. 23.5. Trends of drastic increase or decrease in the uncertainties depict an unhealthy estimation. It is clear from the graphs in Fig. 23.5 that the uncertainties stabilize to small values and do not grow unboundedly.

23.7 Concluding Remarks

The results indicate that EKF estimation provides satisfactory results in identifying coupled dynamics for a small UAV. Especially, for highly noisy input data EKF is

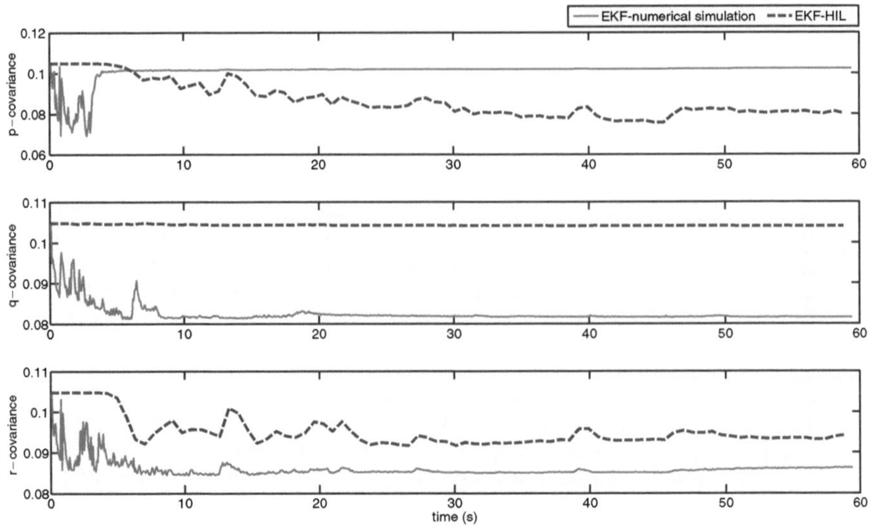

Fig. 23.5. EKF - state covariance analysis

a good option considering its inherent noise filtering properties. Since the computational requirement of EKF is high, it can always be operated at lower sampling rates with reduced updates with acceptable levels of estimation accuracy. These results are also compared with EMId estimation results for a better understanding of these estimation accuracies.

It is shown by means of statistical analysis that the EKF can successfully contain uncertainties in estimation whilst operating with an acceptable percentage of stable sensor data.

Results can be further enhanced by making use of better processors so that EKF can be operated at higher sampling rates with more regular corrections.

23.8 Acknowledgements

We would like to thank Prof Ian Petersen, ITEE, UNSW@ADFA, and Dr Jonghyuk Kim, ANU for their inputs towards this work.

References

1. K.W. Iliff. Parameter estimation for flight vehicles. *Journal of Guidance, Control and Dynamics*, 12:609–622, 1989.
2. M.E. Campbell and S. Brunke. Nonlinear estimation of aircraft models for on-line control customization. In *IEEE Aerospace Conference*, Big Sky MT, 2001.
3. R.V. Jategaonkar and E. Plaetschke. Algorithms for aircraft parameter estimation accounting for process and measurement noise. *Journal of Aircraft*, 26(4):360–372, 1989.

4. R.W. Johnson, S. Jayaram, L. Sun, and J. Zalewski. Distributed processing kalman filter for automated vehicle parameter estimation - a case study. In *Proceedings IASTED Intl Conf. on Applied Simulation and Modeling*, pages 270–276, 2000.
5. R.V. Jategaonkar and E. Plaetschke. Identification of moderately nonlinear flight mechanics systems with additive process and measurement noise. *Journal of Guidance, Control, and Dynamics*, 13(2):277–285, 1990.
6. S.A. Salman, V.R. Puttige, and A.G. Sreenatha. Real-time validation and comparison of fuzzy identification and state-space identification for a uav platform. In *Joint conferences IEEE Conference on Control Applications (CCA), IEEE Computer Aided Control Systems Design Symposium (CACSD) and 2006 IEEE International Symposium on Intelligent Control (ISIC)*, Munich, Germany, 2006.
7. A.G. Kallapur, S.A. Salman, and S.G. Anavatti. Experiences using emid and ekf for uav online identification. In *Third International Conference on Autonomous Robots and Agents*, pages 207 – 212, Palmerston North, New Zealand, 2006.
8. M. Curvo. Estimatin of aircraft aerodynamic derivatives using extended kalman filter. *Journal of The Brazilian Society of Mechanical Sciences*, 22(2):133 – 148, 2000.
9. R.G. Brown and P.Y.C. Hwang. *Introduction to Random Signals and Applied Kalman FIltering*. John Wiley & Sons, Inc., 3rd edition, 1997.
10. S.E. Lyshevski. State-space identification of nonlinear flight dynamics. In *IEEE International Conference on Control Applications*, Hartford, CT, 1997.
11. M.S. Grewal, L.R. Weill, and A.P. Andrews. *Global Positioning Systems, Inertial Navigation, and Integration*. A John Wiley & Sons, Inc., 2001.

24

A Novel Strategy for Multiagent Coalitions in a Dynamic Hostile World

Madhu Goyal

Faculty of Information Technology
University of Technology Sydney, Australia
madhu@it.uts.edu.au

Summary. One of the main underpinning of the multi-agent systems community is how and why autonomous agents should cooperate with one another. Several formal and computational models of cooperative work or coalition are currently developed and used within multi-agent systems research. The coalition facilitates the achievement of cooperation among different agents. In this paper, a mental construct called attitude is proposed and its significance in coalition formation in a dynamic fire world is discussed. It shows that coalitions explore the attitudes and behaviours that help agents to achieve goals that cannot be achieved alone or to maximize net group utility.

Keywords: multi-agent, coalition formation, attitudes.

24.1 Introduction

Coalition formation has been addressed in game theory for some time. However, game theoretic approaches are typically centralized and computationally infeasible. MAS researchers [5–7] using game theory concepts have developed algorithms for coalition formation in MAS environments. However, many of them suffer from a number of important drawbacks like they are only applicable for small number of agents and not applicable to real world domains. A coalition is a group of agents who join together to accomplish a task that requires joint task execution, which otherwise be unable to perform or will perform poorly. It is becoming increasingly important as it increases the ability of agents to execute tasks and maximize their payoffs. Thus the automation of coalition formation will not only save considerable labor time, but also may be more effective at finding beneficial coalitions than human in complex settings. To allow agents to form coalitions, one should devise a coalition formation mechanism that includes a protocol as well as strategies to be implemented by the agents given the protocol.

This chapter will focus on the issues of coalitions in dynamic multiagent systems: specifically, on issues surrounding the formation of coalitions among possibly among heterogeneous group of agents, and on how coalitions adapt to change in

M. Goyal: *A Novel Strategy for Multiagent Coalitions in a Dynamic Hostile World*, Studies in Computational Intelligence (SCI) **76**, 209–215 (2007)
www.springerlink.com

dynamic settings. Traditionally, an agent with complete information can rationalize to form optimal coalitions with its neighbors for problem solving. However, in a noisy and dynamic environment where events occur rapidly, information cannot be relayed among the agent frequently enough, centralized updates and polling are expensive, and the supporting infrastructure may partially fail, agents will be forced to form suboptimal coalitions. In such settings, agents need to reason, with the primary objective of forming a successful coalition rather than an optimal one, and in influencing the coalition (or forming new coalitions) to suit its changing needs. This chapter introduces a novel attitude based coalition agent system in the fire world. The task of fire fighting operations in a highly dynamic and hostile environment is a challenging problem. We suggest a knowledge-based approach to the coalition formation problem for fire fighting missions. Owing to the special nature of this domain, developing a protocol that enables agents to negotiate and form coalitions and provide them with simple heuristics for choosing coalition partners is quite challenging task.

24.2 A Fire Fighting World

We have implemented our formalization on a simulation of fire world ([4]) using a virtual research campus. The fire world is a dynamic, distributed, interactive, simulated fire environment where agents are working together to solve problems, for example, rescuing victims and extinguishing fire. Humans and animals in the fire world are modeled as autonomous and heterogeneous agents. While the animals run away from fire instinctively, the fire fighters can tackle and extinguish fire, and the victims escape from fire in an intelligent fashion. An agent responds to fire at different levels. At the lower level, the agent burns like any object, such as chair. At the higher level, the agent reacts to fire by quickly performing actions, generating goals and achieving goals through plan execution. Agents operating in the domain face a high level of uncertainty caused by the fire. There are three main objectives for intelligent agents in the world during the event of fire: self-survival, saving objects including lives of animals and other agents, and put-off fire. Because of the hostile settings of the domain, there exist a lot of challenging situations where agents need to do the cooperative activities. Whenever there is fire, there is need of coalition between the fire fighters (FF-agent), volunteers (Vol-agent), and victim agents (Vic-agent) (Fig. 24.1). The fire fighters perform all the tasks necessary to control an emergency scene. The problem solving activities of the fire fighters are putting out fire, rescuing victims, and saving property. Apart from these primary activities there are a number of subtasks e.g., run toward the exit, move the objects out of the room, remove obstacles, and to prevent the spread of fire. The first and paramount objective of the victim agents is self-survival. The role of volunteer agents is to try to save objects from the fire and help out other victims who need assistance when they believe their lives are not under threat. To achieve these tasks there is need of coalitions between these agents is necessary. Thus the fire world we consider is sufficiently complex to bring about the challenges involved in to study coalition formation in a typical multiagent world.

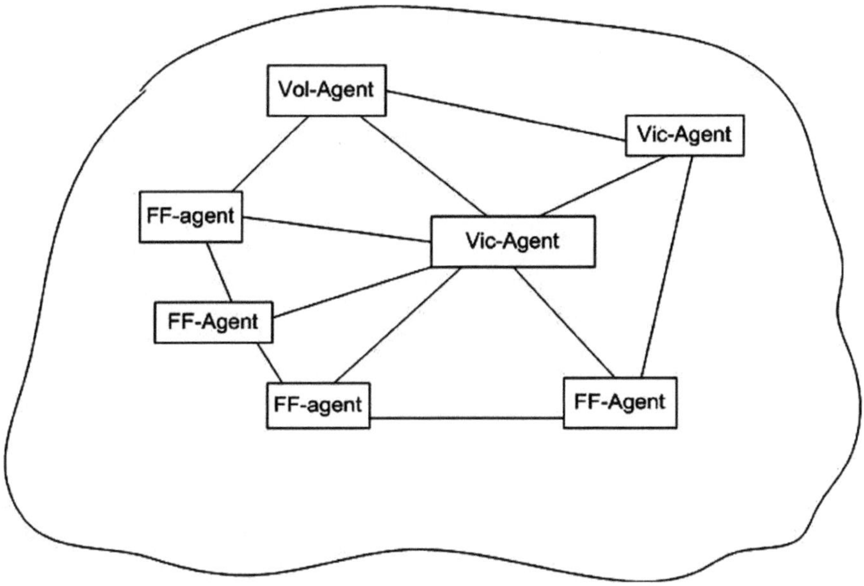

Fig. 24.1. Coalition between fire-fighter, volunteer, and victim agent

24.3 Strategic Coalitions in an Agent-Based Hostile World

The coalition facilitates the achievement of cooperation among different agents. The cooperation among agents succeeds only when participating agents are enthusiastically unified in pursuit of a common objective rather than individual agendas. We claim that cooperation among agents is achieved only if the agents have a *collective attitude* toward cooperative goal as well as toward cooperative plan. From collective attitudes, agents derive individual attitudes that are then used to guide their behaviors to achieve the coalition activity. The agents in a coalition can have different attitudes depending upon the type of the environment the agent occupies.

24.3.1 Definition of Attitude

Attitude is a *learned predisposition to respond in a consistently favorable or unfavorable manner with respect to a given object* ([3]). In other words, the attitude is a preparation in advance of the actual response, constitutes an important determinant of the ensuing behavior. However this definition seems too abstract for computational purposes. In AI, the fundamental notions to generate the desirable behaviors of the agents often include goals, beliefs, intentions, and commitments. Bratman ([1]) addresses the problem of defining the nature of intentions. Crucial to his argument is the subtle distinction between doing something intentionally and intending to do something. Cohen and Levesque ([2]), on the other hand, developed a logic in which intention is defined. They define the notion of individual commitment as *persistent goal*, and an intention is defined to be a commitment to act in a certain mental state of believing throughout what he is doing. Thus to provide a definition of attitude that is concrete enough for computational purposes, we model attitude using goals, beliefs, intentions, and commitments. From the Fishbein's definition it is clear that when an attitude is adopted, an agent has to exhibit an appropriate behavior (predisposition means behave in a particular way).

In a dynamic multiagent world, the behavior is also based on appropriate commitment of the agent to all unexpected situations in the world including state changes, failures, and other agents' mental and physical behaviors. An agent intending to achieve a goal must first commit itself to the goal by assigning the necessary resources, and then carry out the commitment when the appropriate opportune comes. Second, if the agent is committed to executing its action, it needs to know how weak or strong the commitment is. If the commitment is weak, the agent may not want to expend too much of its resources in achieving the execution. The agent thus needs to know the degree of its commitment toward the action. This degree of commitment quantifies the agent's *attitude* toward the action execution. For example, if the agent considers the action execution to be *higher importance* (an attitude toward the action), then it may choose to execute the action with greater degree of commitment; otherwise, the agent may drop the action even when it had failed at the first time. Thus, in our formulation, an agent when it performs an activity, since the activity is more likely that it will not succeed in a dynamic world; agents will adopt a definite attitude toward every activity while performing that activity. The adopted attitude will *guide* the agent in responding to failure situations. Also the behavior must be consistent over the period of time during which the agent is holding the attitude. An agent cannot thus afford to change its attitude toward a given object too often, because if it does, its behavior will become somewhat like a reactive agent, and its attitude may not be useful to other agents. Once an agent chose to adopt an attitude, it strives to maintain this attitude, until it reaches a situation where the agent may choose to drop its current attitude toward the object and adopt a new attitude toward the same object. Thus we define attitude as: *An agent's attitude toward an object is its persistent degree of commitment to one or several goals associated with the object, which give rise to persistent favorable or unfavorable behavior to do some physical or mental actions.*

24.3.2 Type of Attitudes

Behaviors exhibited by an agent in a multiagent environment can be either individualistic or collective. Accordingly, we can divide attitudes in two broad categories: *individual* attitudes and *collective* attitudes. The *individual* attitudes contribute toward the single agent's view toward an object or person. The agent's individual attitude toward a fire world, for example, is a function of its beliefs about the fire world. The *collective* attitudes are those attitudes, which are held by multiple agents. The collective attitudes are individual attitudes so strongly interconditioned by collective contact that they become highly standardised and uniform within the group, team, society, etc. The agents can collectively exist as *societies, groups, teams, friends, foes*, or just as *strangers*, and *collective* attitudes are possible in any one of these classifications. For example, the agents in the collection called *friends* can all have a collective attitude called *friends*, which is mutually believed by all agents in the collection. A collective attitude can be viewed as an abstract attitude consisting of several component attitudes, and for an individual agent to perform an appropriate behavior; it must hold its own attitude toward the collective attitude. Thus, for example, if A_1 and A_2 are *friends*, then they mutually believe they are friends, but also each A_i must have an attitude toward this infinite nesting of beliefs so that it can exhibit a corresponding behavior. Thus, from A_1's viewpoint, *friends* is an attitude that it is holding toward the collection $\{A_1, A_2\}$ and can be denoted as $friendsA_1$ (A_1, A_2). Similarly, from A_2's view point, its attitude can be denoted as $friendsA_2$ (A_1, A_2). However, in an extreme case, A1 may not be certain about A_2's behavior. That is why; it needs to have an attitude of its own, which generates a behavior taking all the uncertainties introduced by the dynamic environment into account. Further note that A_1 might implicitly expect A_2 to perform its role, but it is only an expectation. This is a bottom up view of the friends-relation, where the relation is viewed in general differently by each agent depending on the local situations the individual agents face in the world.

24.3.3 Attitude-Based Coalition Model

We claim that successful coalition is achieved only if the agents have coalition as a collective abstract attitude. From this collective attitude, agents derive individual attitudes that are then used to guide their behaviors to achieve the coalition. Suppose there n agents in a coalition i.e., $A_1 \ldots . A_n$. So the collective attitude of the agent $A_1 \ldots . A_n$ toward the coalition is represented as $Coal_{A1..An}(A_1, \ldots, A_n)$. But from A_1's viewpoint, team is an attitude that it is holding toward the collection $(A_1, \ldots, A_n)$ and can be denoted as $Coal_{A1}(A_1, A_2)$. Similarly from A_n's viewpoint, its attitude can be denoted as $Coal_{An}(A_1, \ldots, A_n)$. But the collective attitude $Coal_{A1An}(A_1, \ldots, A_n)$ is decomposed into the individual attitudes only when all the agents mutually believe that they are in the coalition. The coalition attitude can be represented in the form of individual attitudes toward the various attributes of the coalition i.e., coalition methods, coalition rule base, and coalition responsibility. The attitudes of an agent existing in a coalition consist of attitude toward coalition as well as attitude toward coalition activity. At any time, an agent may be engaged in one of the basic coalition activities i.e., coalition formation, coalition maintenance, and coalition dissolution. Instead of modeling these basic activities as tasks to be achieved, we have chosen to model them as attitudes.

Coalition (A1,..,A2)

This attitude is invoked when the agents are in a team state. This attitude guides the agents to perform the appropriate coalition behaviors.

Name of Attitude: ***Coalition***

Description of Object: (1) Name of Object: set of agents (2) Model of Object: $\{A_1, A_n | A_i \text{ is an agent}\}$

Basic agent behavior: coalition behavior specified by agent's rule base

Evaluation: favorable

Persistence: This attitude persists as long as the agents are able to maintain it.

Concurrent attitudes: all attitudes toward physical and mental objects in the domain.

Type of Attitude: collective.

Coalition Formation

Impetus for attitude and coalition formation may arise from the world and a particular domain or from agent's themselves. Having identified the potential for coalition action with respect to one of its goals, a leader agent will solicit assistance from some group of agents that it believes can achieve the goal. Our agents form coalition, because the inherent nature of the world requires agents to exist and act together. However, a particular situation may force the agents to dissolve the coalition for some time. In the fire world, the event triggering the coalition formation process is a fire. Whenever there is fire, the security officers call the fire-fighting company to put out the fire. Then the fire fighters arrive at the scene of fire and get the information about when, how, and where the fire had started. Suppose there is a medium fire in the campus, which results in the attitudes *medium-fire* and *dangerous-fire* toward the object fire. The attitude *Coal-form* is also generated, which initiates the team formation process. We propose a dynamic team formation model, in which we consider initially the mental state i.e., the beliefs of all the agents is same. The fire-fighting agents recognize appropriateness of the team model for the task at hand; set up the requirements in terms of other fellow agents, role designation, and structure; and develop attitudes toward the team as well as toward the domain. In order to select a member of the team, our agent will select the fellow agent who has following capabilities:

- Has knowledge about the state of other agents
- Has attitude towards the coalition formation
- Can derive roles for other agents based on skills and capabilities
- Can derive a complete joint plan
- Can maintain a coalition state

Our method of forming a coalition is like this; the agents start broadcasting message to other agents "Let us form a coalition." The agents will form a coalition if two or more than two agents agree by saying, "Yes." If the agent do not receive the "Yes" message, it will again iterate through the same steps until the coalition is formed. The *coal-form* is maintained as long as the agents are forming the team. Once the team is formed, agents will drop the *coal-form* attitude and form the *coal* attitude, which will guide the agents to produce various team behaviors.

Coalition Maintenance and Dissolution

While solving a problem (during fire fighting activity) the coalition agents have also to maintain the coalition. During the coalition activity the agents implement the coalition plan to achieve the desired coalition action and sustain the desired consequences. The coalition maintenance behavior requires what the agent should do so that coalition does not disintegrate. In order to maintain the coalition each agent should ask the other agent periodically or whenever there is a change in the world state, whether he is in the coalition. So the attitudes like *periodic–coalition–maintenance* and *situation–coalition–maintenance* are produced periodically or whenever there is a change in the situation. These attitudes help the agent to exhibit the maintenance behaviors. When the team task is achieved or team activity has to be stopped due to unavoidable circumstances, the attitude *coal-unform* is generated. This attitude results in the dissolution of the team and further generates attitude *escape*. For example, when the fire becomes very large, the agents have to abandon the team activity and escape. The attitude *coal–unform* is maintained as long as the agents are escaping to a safe place. Once the agents are in the safe place, the attitudes *team-unform* and *escape* are relinquished. In case the fire comes under control, the agents again form a team by going through the steps of team formation.

24.4 Conclusions

This chapter has developed a novel framework for managing coalitions in a hostile dynamic world. Coalition is guided by the agent's dynamic assessment of agent's attitudes given the current scenario conditions, with the aim of facilitating the agents in coalitions to complete their tasks as quickly as possible. In particular, it is outlined in this paper that how agents can form and maintain a coalition, and how it can offer certain benefits to cooperation. Our solution provides a means of maximizing the utility and predictability of the agents as a whole. Its richness presents numerous possibilities for studying different patterns of collaborative behavior.

References

1. Bratman ME (1987) Intentions, Plans and Practical Reason. Harvard University Press, Cambridge, MA.
2. Cohen PR, Levesque HJ (1991) Teamwork, Special Issue on Cognitive Science and Artificial Intelligence, 25(4).
3. Fishbein M, Ajzen I (1975) Belief, Attitude, Intention and Behaviour: An Introduction to theory and research. Reading, MA, USA: Addison-Wesley.
4. Goyal M (2006) Decision making in Multi-agent Coalitions in a Dynamic Hostile World, The Third International Conference on Autonomous Robots and Agents, Palmerston North, New Zealand.
5. Griffiths N, Luck M (2003) Coalition Formation Through Motivation and Trust, Proceedings of the 2nd International Conference on Autonomous Agents and Multi-agent Systems, Melbourne, Australia, pp 17–24, ACM Press.
6. Kraus S, Shehory O, Taase G (2003) Coalition formation with uncertain heterogeneous information, Proceedings of the 2nd International Conference on Autonomous Agents and Multi-agent Systems, Melbourne, Australia.
7. Sandholm T, Larson K, Andersson M, Shehory O, Tohmé F (1999) Coalition Structure Generation with Worst Case Guarantees. Artificial Intelligence, 111(1–2), pp 209–238.

25

Emotion Recognition Using Voice Based on Emotion-Sensitive Frequency Ranges

Kyung Hak Hyun, Eun Ho Kim, and Yoon Keun Kwak

Department of Mechanical Engineering
Korea Advanced Institute of Science and Technology
Daejeon, Republic of Korea
cromno9@kaist.ac.kr, kimeunho@kaist.ac.kr, ykkwak@kaist.ac.kr

Summary. To date, study on emotion recognition has focused on detecting the values of pitch, formant, or cepstrum from the variation of speech according to changing emotions. However, the values of emotional speech features vary by not only emotions but also speakers. Because each speaker has unique frequency characteristics, it is difficult to apply the same manner to different speakers. Therefore, in the present work we considered the personal characteristics of speech. To this end, we analyzed the frequency characteristics for a user and chose the frequency ranges that are sensitive to variation of emotion. From these results, we designed a personal filter bank and extracted emotional speech features using this filter bank. This method showed about 90% recognition rate although there are differences among individuals.

Keywords: Emotion, Recognition, Frequency range, Filter.

25.1 Introduction

Emotional speech recognition involves automatically identifying the emotional or physical state of a human being from voice. The importance of emotion recognition for human computer interaction is widely recognized [9]. Although the emotional state does not alter linguistic content, it is an important factor in human communication and it also provides feedback information in many applications. In human–machine interactions, the machine can be made to produce more appropriate responses if the state of emotion of the person is accurately identified. Other applications of an automatic emotion recognition system include tutoring, alerting, and entertainment [3].

25.1.1 Previous Works

The first investigations for human emotions were conducted in the mid-1980s using statistical properties of certain acoustic features [2, 13]. In 2000, emotional speech

K.H. Hyun et al.: *Emotion Recognition Using Voice Based on Emotion-Sensitive Frequency Ranges*, Studies in Computational Intelligence (SCI) **76**, 217–223 (2007)
www.springerlink.com

recognition was employed by therapists as a diagnostic tool in medicine [5]. Presently, most researches are focused on finding powerful combinations of classifiers that advance the classification efficiency in real-life applications. As an example, ticket reservations, so-called "SmartKom," can recognize customer frustration and change their response accordingly [1, 12].

A correlation of emotion and frequency was usually used for emotional speech features but the feature characteristic was changed a lot by different speaker. Hence, the results of emotion recognition in speaker independent systems are below 60% [4, 7].

Therefore, in this study, we analyzed the frequency characteristics of speakers and applied a personal filter bank, which was proposed in [6].

25.2 Emotional Speech Database

The Korean database was used in our experiment. The database was recorded in the framework of the G7 project in Korea. In this database emotional sentences from five male speakers and five female speakers were recorded.

25.2.1 Corpus of Database

Each corpus contains several phonemes. The database contains 45 sentences. The corpora contain short, medium, and long sentences that are context independent. The sentences are chosen upon consideration of the followings:

1. The sentence is able to pronounce in several emotional states.
2. The sentence can express the emotions naturally.
3. The database should contain all phonemes of Korean.
4. The database should contain several dictions such as honorific words.

25.2.2 Speech Materials

The corpora were recorded over four basic emotions. The recorded emotions are neutrality (N), joy (J), sadness (S), and anger (A).

The database contains speech recorded in three iterations. All speech was recorded four times and the worst record among those was discarded. The aim of this step is to filter clumsy wording and maintain consistency of the speech.

The database contains 5,400 sentences. The subjective evaluation tests for all corpora were performed.

25.2.3 Subjective Evaluation Test

Subjective evaluation tests were made for the database. The subjective evaluation test included 30 listeners. The listeners were engineering students from Yonsei University in Korea. Each listener decided which emotion corresponded to each utterance. The samples were played randomly. Ten listeners made decisions for each utterance. The results of the subjective evaluation test showed, on average, 78.2% accuracy. The confusion matrix of the test is shown in Table 25.1.

Table 25.1. Human performance

Recog. (%)	Neutrality	Joy	Sadness	Anger
Neutrality	83.9	3.1	8.9	4.1
Joy	26.6	57.8	3.5	12.0
Sadness	6.4	0.6	92.2	0.8
Anger	15.1	5.4	1.0	78.5
Overall			78.2	

25.3 Personal Filter Bank Design

25.3.1 Log Frequency Power Coefficients

Humans perceive audible sound from 20 Hz to 20 kHz. Furthermore, human perception is not linear to physical frequency, and hence there is a different unit for audible sound frequency, that is, Mel [10].

The human auditory system has several sensitive frequency ranges. Under 1,000 Hz, humans feel the pitch is linear to physical frequency; however, when the frequency is increased above 1,000 Hz, the auditory system becomes insensitive. In other words, as frequency changes become larger, humans can distinguish changes of pitch.

Accordingly, we can regard the human audible system as a filter bank, as described in [11]. Therefore, a log frequency filter bank can be regarded as a model that follows the varying auditory resolving power of the human ear for various frequencies.

In Tin Lay Nwe's study [8], a filter bank is designed to divide speech signals into 12 frequency bands that match the critical perceptual bands of the human ear. The center frequencies and bandwidths of 12 filter banks, which were proposed by Tin Lay Nwe, are described in [8].

25.3.2 Filter Bank Analysis

In the analysis of frequency characteristics of a user, we attempted to identify specific frequency ranges that are sensitive to changes of emotion and robust to changes of context. To this end, we compared the rates of emotional speech recognition by changing the filter bank.

The results are presented in Figs. 25.1 and 25.2. Both show normalized values on the y-axis in a range between 0 and 1 to verify the effect of each filter. In Fig. 25.1, the recognition rates are for speech that was filtered by a one order filter. Hence, we can estimate which filter is most useful with respect to recognizing the emotion. Figure 25.2 shows the results for speech that was filtered by an eleven order filter bank. Thus, we can identify the worst filter.

It is possible to select the best filter or abandon the worst filter on the basis of the above results. In this regard, the results demonstrate that accepting or rejecting a specific filter from the filter bank can affect the emotional recognition results.

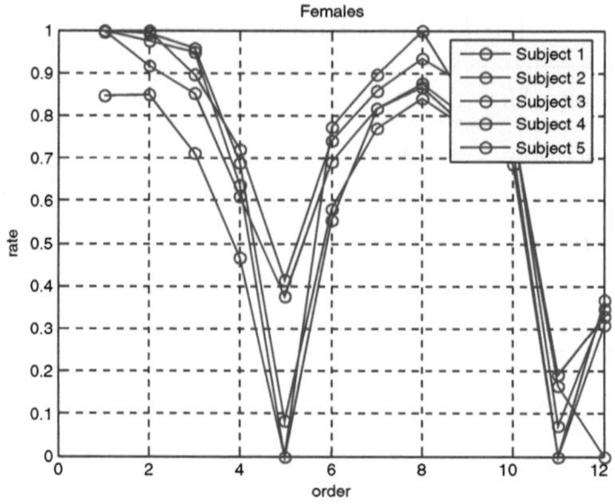

Fig. 25.1. One filter selection in female subjects

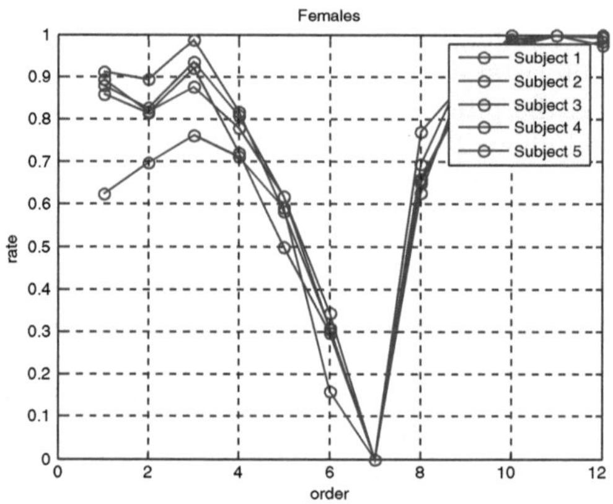

Fig. 25.2. One filter exception in female subjects

However, they do not indicate the best set of filters to recognize the emotion, because each filter of the filter bank is not independent of the other filters. According to the study of Rabiner and Juang [11] and a hypothesis of Tin Lay Nwe, each filter shares frequency ranges with its neighbor filters. Therefore, a filter cannot be chosen independently from the filter bank and should be chosen under consideration of the effects of other filters.

25.3.3 Design of the Personal Filter Bank

Design Method

From the analysis presented above, we find that there are different frequency ranges that can help in the recognition of emotion in speech. Hence, we compared the

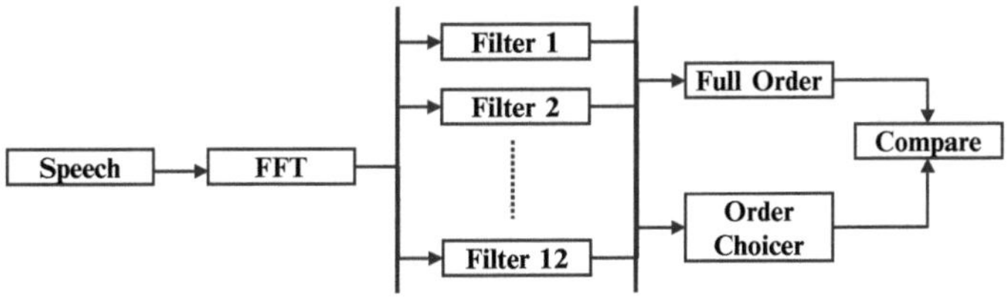

Fig. 25.3. Flow chart of the adaptive filter bank method

Table 25.2. Adaptive filter orders and the improvement rate

Subject	Adaptive filter order	Improvement (%)
Female 1	3,4,5,7,9,10,11,12	97
Female 2	2,5,6,7,8,9,10,11,12	93
Female 3	4,6,7,9,11,12	98
Female 4	1,4,5,6,8,10,11	94
Female 5	3,5,6,7,8,9,10,11,12	84

results obtained using a full order filter bank with those derived using a selected filter set.

The full order filter bank was selected from Tin Lay Nwe's study [8]. For comparison, we assessed several filter sets in order to identify the best set in terms of performance. At this step the full order is twelve and each filter can be selected independently. Thus, it is necessary to compare 4 095 cases ($2^{12} - 1$) for one person.

Finally, we examined the results obtained using a Bayesian Classifier and chose the best set for each person. A summary of this procedure is presented in Fig. 25.3. In this figure, full order means the classifier uses the full information from filters $1 \sim 12$, whereas order choicer means the classifier adaptively uses the information among the filters.

Evaluation

To evaluate the designed filter bank, we compared the results of the designed filter bank with those of the full-order filter bank. We performed 100 iterations for the designed filter bank for cross-validation in order to verify the results statistically. For the cross-validation, the training and test data were randomly chosen from the database; 120 training data were chosen randomly and 300 test data were chosen not only randomly but also differently from the training data for the evaluation.

We found that the designed filter bank method sometimes decreased the recognition rate. However, the degree of decrease was very small. From Table 25.2, however, $80 \sim 90\%$ of cases showed an increase in the recognition rate for 100 repetitions of the experiment for each subject.

Finally, the recognition rates of all subjects are presented in Fig. 25.4. In addition, we present a comparison with the principal component analysis (PCA) results.

Hence, we find that the proposed method is superior to the PCA method, as the latter can only find the principal component, which is good for representation, whereas the proposed method finds the frequency ranges that are sensitive to change of emotion.

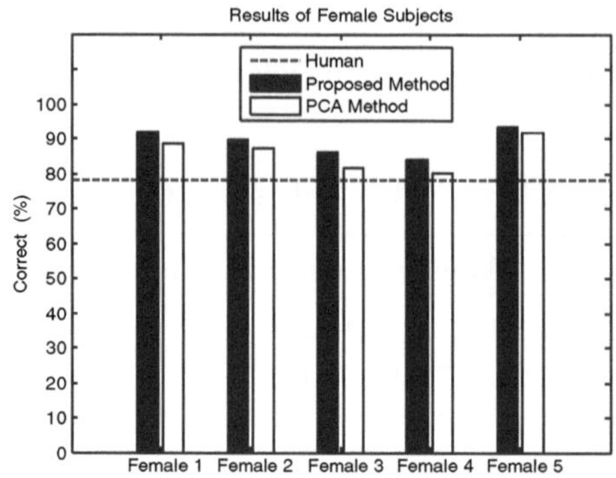

Fig. 25.4. Result of female subjects

Discussion

Although proposed method is available with learning data for each subject and collecting user's speech data with emotion is not easy, this works contributed that each person has different frequency characteristics in emotional speech and some frequency ranges is not important to classify emotions. In proposed method, we analyzed each subject's frequency characteristics by filter bank method and designed personal filter bank for emotion recognition. As presented in Fig. 25.4, proposed method shows better result than PCA method and human performance. In simulation, we assumed clean environment because all speech data were loaded from database; however, the noise problem is serious in real application. Proposed method generally uses high frequency regions which are easily distorted by noise. Therefore, noise reduction process will be needed before applying proposed method to attain the desired results.

25.4 Conclusion

In this work, we consider the correlation of speech frequency with emotion and develop a personal filter bank for emotion recognition. We verified the superiority of the proposed method through comparison with PCA results. The strength of the proposed approach lies in selecting frequency ranges that are sensitive to the change of emotion. Although the improvement varied on a case by case and person to person basis, a roughly $2 \sim 5\%$ overall improvement in recognition rate was obtained.

References

1. Ang J, Dhillon R, Krupski A, Shriberg E, Stolcke A (2002) Prosody-based automatic detection of annoyance and frustration in human-computer dialog. In: Hansen JHL, Pellom B (eds) International Conference on Spoken Language Processing, ISCA Archive, pp 2037–2040
2. Bezooijen Rv (1984) The Characteristics and Recognizability of Vocal Expression of Emotions. Walter de Gruyter, Inc., The Netherlands
3. Cowie R, Douglas-Cowie E, Tsapatsoulis N, Votsis G, Kollias S, Fellenz W, Taylor JG (2001) Emotion recognition in human-computer interaction. In: Chang S-F, Schneiderman S, (eds) IEEE Signal Processing Magazine, IEEE Signal Processing Society, pp 32–80
4. Esau N, Kleinjohann B, Kleinjohann L, Stichling D (2003) MEXI: Machine with Emotionally eXtended Intelligence. In: Abraham A, Köppen M, Franke K (eds) Hybrid Intelligent systems, Design and Application, IOS Press, The Netherlands pp 961–970
5. France DJ, Shivavi RG, Silverman S, Silverman M, Wilkes M (2000) Acoustical properties of speech as indicators of depression and suicidal risk. IEEE Trans Biomed Eng 7:829–837
6. Hyun KH, Kim EH, Kwak YK (2006) Emotion Recognition Using Frequency Ranges Sensitive to Emotion In: Proceeding of 3rd International Conference on Autonomous Robots and Agents, pp 119–124
7. McGilloway S, Cowie R, Douglas-Cowie E, Gielen S, Westerdijk M, Stroeve S (2000) Approaching automatic recognition of emotion from voice: A rough benchmark. In: Proceeding of the ISCA Workshop on Speech and Emotion, pp 207–212
8. Nwe TL, Foo SW, De Silva LC (2003) Speech emotion recognition using hidden markov model. Speech Communication, 41:603–623.
9. Pantic M, Rothkrantz L (2003) Toward an affect-sensitive multimodal human-computer interaction. In: Trew RJ, Calder J (eds) Proceeding of the IEEE, IEEE, pp 1370–1390
10. Quatieri TF (2002) Discrete-Time Speech Signal Processing Principles and Practice, Prentice Hall, New Jersey.
11. Rabiner LR, Juang BH (1993) Fundamentals of Speech Recognition, Prentice Hall, Englewood Cliffs, New Jersey
12. Schiel F, Steininger S, Turk U (2002) The Smartkom multimodal copus at BAS. In: The 3rd International Conference on Language Resources and Evaluation, pp 35–41
13. Tolkmitt FJ, Scherer KR (1986) Effect of experimentally induced stress on vocal parameters. J Exp Psychol: Hum Percept Perform 12:302–313

References

26

FPGA-Based Implementation of Graph Colouring Algorithms

Valery Sklyarov, Iouliia Skliarova, and Bruno Pimentel

Department of Electronics, Telecommunications and Informatics
IEETA University of Aveiro, Aveiro, Portugal
skl@det.ua.pt, iouliia@det.ua.pt, pimentel@ieeta.pt

Summary. The chapter suggests an FPGA-based implementation of graph colouring algorithms formulated over ternary matrices. First, software models (described in C + +) and the respective hardware circuits (synthesised from Handel-C specifications) for getting exact solutions are discussed, analysed and compared. Then it is shown that the exact algorithm can serve as a base for a number of approximate algorithms, which permit to improve incrementally the results until some predefined criteria are satisfied. Characteristics and capabilities of the approximate algorithms are also examined.

Keywords: Graph colouring, Discrete matrices, FPGA implementation.

26.1 Introduction

For solving the graph-colouring problem it is required to paint graph vertices in such a way that any two vertices connected with an edge are painted in different colours and the number of used colours is minimal. Graph colouring algorithms are widely used for solving different engineering problems in robotics and embedded systems [2–4, 11, 12].

It should be noted that the graph colouring is a very computationally complex task (this is NP-hard problem [7]). Often it is necessary to solve this problem in run-time (for example, for register allocation [4]), which increases essentially the total execution time for the relevant applications. However, acceleration can be achieved with the aid of graph-colouring targeted hardware [5, 6]. This makes possible the number of the required clock cycles for the respective applications to be decreased significantly.

The chapter suggests algorithms for solving the graph colouring problem in FPGA. In order to find out a compromise between the execution time and the required accuracy two types of algorithms have been implemented and analysed. They are an exact algorithm producing the optimal solution and approximate algorithms enabling us to stop the execution as soon as an appropriate for the considered

V. Sklyarov et al.: *FPGA-Based Implementation of Graph Colouring Algorithms*, Studies in Computational Intelligence (SCI) **76**, 225–231 (2007)
www.springerlink.com

problem solution has been found. The considered materials are derived from [10], where they were first presented and discussed.

26.2 Matrix Specification

Binary and ternary matrices are very well suited for processing them in FPGA [8]. This section demonstrates that any graph, which has to be coloured, can be converted to matrix specification in such a way that solving the problem over the matrix is equivalent to solving the problem over the graph. The graph [14] in Fig. 26.1 will be used as an example.

Let us consider a matrix, which has the same number of rows as the number N of vertices of the graph G. If and only if two vertices m_i and m_j are connected with an edge in G then the matrix rows i and j must be orthogonal. Orthogonality of two ternary vectors is defined as follows [14]: $(m_i \text{ ort } m_j) \Rightarrow \{m_i\} \cap \{m_j\} = \emptyset$, where $\{m_i\}$ ($\{m_j\}$) is a set of binary vectors that correspond to the ternary vector m_i (m_j) by replacing the don't care values (−) with all possible combinations of 0s and 1s. For example, the following two ternary vectors 0–11- and -1110 are not orthogonal, because $\{00110,00111,01110,01111\} \cap \{01110,11110\} = \{01110\} \neq \emptyset$. If and only if two ternary vectors are not orthogonal (let us designate this as $m_i \not{ort}\, m_j$) they do intersect in the Boolean space: $(m_i \not{ort}\, m_j) \Leftrightarrow (m_i \text{ ins } m_j)$.

Table 26.1 depicts matrix $\mu(G)$ that was built for the graph G in Fig. 26.1. This is a symmetric matrix with respect to the main diagonal that is why just a lower triangle of the matrix is sufficient.

The matrix was built using the following rules:

1. Provide orthogonality of the 1st vertex with the vertices connected to the 1st vertex by edges. For our example they are 2, 6, 8 and 10 (see Fig. 26.1). This can be done by recording the value '0' in the first column (I) of the row 1 and the value '1' in the first column (I) of the rows 2, 6, 8 and 10 of Table 26.1. All empty elements of the first column are assigned to *don't care* ('–'). Correctness of these operations is evident from the definition of orthogonality.
2. If the vertex 2 has connections with succeeding vertices, i.e. with the vertices 3, 4, 5,...then use column II to indicate orthogonality; otherwise, skip vertex

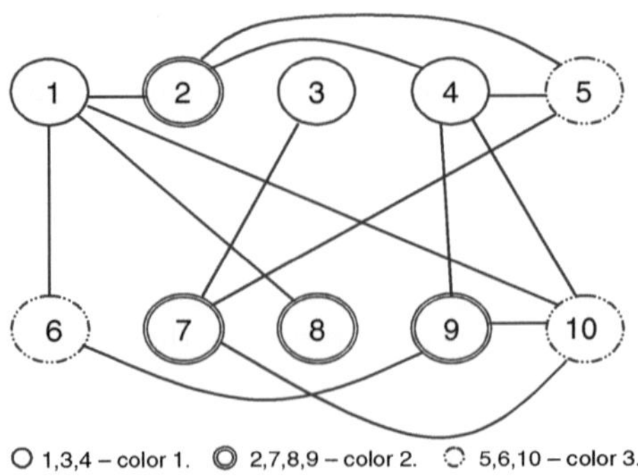

Fig. 26.1. Graph G and its optimal colouring

Table 26.1. Matrix μ(G) with rows $1, 2, \ldots, 10$ and columns I, II, …, VIII

	I	II	III	IV	V	VI	VII	VIII
1	0	–	–	–	–	–	–	–
2	1	0	–	–	–	–	–	–
3	–	–	0	–	–	–	–	–
4	–	1	–	0	–	–	–	–
5	–	1	–	1	0	–	–	–
6	1	–	–	–	–	0	–	–
7	–	–	1	–	1	–	0	–
8	1	–	–	–	–	–	–	–
9	–	–	–	1	–	1	–	0
10	1	–	–	1	–	–	1	1

2 and consider the next one. For our example the vertex 2 is connected with the succeeding vertices 4 and 5. Thus, the column II contains '1's in the rows 4 and 5 and '0' in the row 2. As in the previous case, all the remaining elements of column II receive *don't care* values.

3. Use the same rules for all the remaining vertices. For our example it permits to fill in Table 26.1.

Obviously, the maximum number of columns for the constructed matrix μ(G) cannot exceed the value N−1, where N is the number of vertices in the graph G. After getting the matrix, it is necessary to find out the minimal number K of such subsets of rows that any subset does not contain mutually orthogonal rows. In other words, any two of the rows belonging to the same subset must intersect in the Boolean space. The number K will be the minimal number of colours for graph G and rows from each subset will correspond to the vertices of graph G that can be painted in the same colour.

26.3 The Exact Algorithm

The considered exact backtracking search algorithm involves traditional reduction (allowing the matrix to be simplified by deleting some rows and columns) and splitting (permitting the problem to be decomposed in less complicated subproblems) steps that are common to combinatorial search methods [14] and that are repeated sequentially until the solution is found.

Some of the reduction and splitting rules can be directly borrowed from the method of condensation, proposed in [14]. The following reduction rules can be used:

1. If, after selecting a new colour, the matrix μ(G) contains a column i without values '0' ('1') then this column i can be deleted
2. If, at any intermediate step of the algorithm, the matrix μ(G) contains a row j with just don't cares ('–') then this row j can be deleted from the matrix and included in the constructed subset

3. All the rows that have already been included in the constructed subsets are removed from the matrix $\mu(G)$.

The proposed algorithm includes the following steps:

1. Choose a new colour (i.e. create a new initially empty subset)
2. Apply the reduction rules
3. Consider the topmost row m_i in the matrix
4. Include the row m_i in the constructed subset and delete it from the matrix
5. Find out all other rows intersecting with the vector m_i
6. Select the first not tried yet row m_j from point 5, include it in the constructed subset and then delete it from the matrix
7. Assign $m_i = m_i \; ins \; m_j$ and repeat the steps 5–7 if this is possible. If this is not possible go to the step 8
8. If the intermediate matrix is not empty repeat the steps 1–7. Otherwise, store the solution found and then backtrack to the nearest branching point (set at the step 6 and try to find a better solution by repeating the steps 6–8

26.4 Derived Approximate Algorithms

The considered algorithm possesses the following primary advantage. Beginning from the first iteration we will get a solution of the problem. As a rule, this solution is not far away from the optimal solution and therefore might be appropriate for some practical applications. It is important that any new iteration does not make worse the first result and it can only improve it. Thus, either the number of iterations might be limited or any intermediate complete result can be checked for adequacy. This permits to suggest a number of approximate algorithms derived from the exact algorithm. It is very interesting to estimate how the optimal result depends on the number of iterations; how far is the first result from the optimal result, etc. The details of relevant experiments will be reported in Sect. 26.6.

26.5 General Description of FPGA-Targeted Projects

Initially, the exact algorithm for graph colouring was described in C + + language and carefully debugged and tested in PC. Then C + + code was converted to a Handel-C project using the rules established by *Celoxica* [1]. The basic blocks of the hardware project are shown in Fig. 26.2.

The initial and all the intermediate matrices are kept in the same RAM (constructed from embedded in FPGA memory blocks). Mask registers (for columns and for rows) make possible some rows and columns to be excluded from the matrix in order to select a minor (i.e. to specify the reduced matrix). The column/row address registers provide dual access allowing either a row or a column to be read (to a column/row register). The sequence of the required operations is generated by a hierarchical finite state machine [9].

It is important to note that all the blocks in Fig. 26.2 are parameterizable. This property allows for scalability in such a way that the considered blocks can be applicable to matrices of different dimensions.

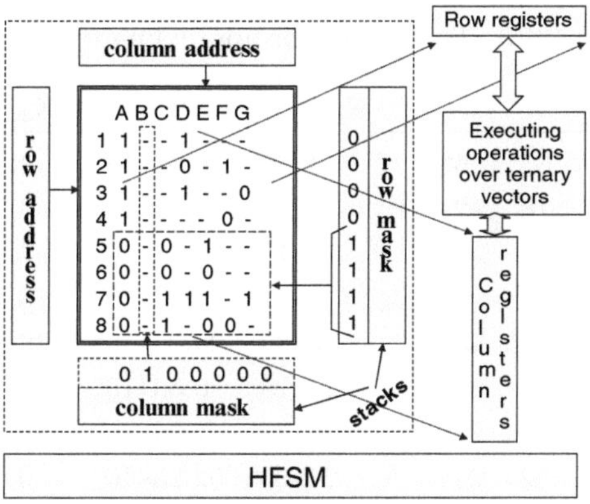

Fig. 26.2. General structure of the hardware project for solving the graph colouring problem

26.6 Experiments and Comparison

The synthesis and implementation of circuits from the specification in Handel-C was done in Celoxica DK4 design suite [1] and implementation was finalised in Xilinx ISE 8.1 for xc2v1000-4fg456 FPGA (Virtex-II family [13]) available on RC200 prototyping board [1]. Note that each circuit includes not only components that are needed for problem analysis, but also auxiliary blocks for entering input data and visualising the results.

The results of experiments are presented in Table 26.2 for the exact algorithm and in Table 26.3 for approximate algorithms.

Rows/columns of the tables contain the following data: P(K) – the number P of problem instance, which can be optimally painted with K colours; N – the number of graph vertices; L – the number of graph edges; N_s – the number of occupied FPGA slices; F_h – the resulting circuit clock frequency in MHz; $N_{clk}{}^s$ – the number of clock cycles required for solving a given problem instance in software; $N_{clk}{}^h$ – the number of clock cycles required for solving a given problem instance in hardware; K_a – the number of colours obtained by the approximate algorithm; and R – the number of respective iterations for K_a colours.

Table 26.2. The results for the exact algorithm

P(K)	N	L	N_s	F_h	$N_{clk}{}^h$	$N_{clk}{}^s$
1(4)	32	53	3481	26	587103122	$7.3568*10^9$
2(3)	18	21	4028	26	166755	$3.99*10^6$
3(4)	28	50	4013	26	102509429	$1.77*10^9$
4(3)	18	12	4028	26	24122	$0.426*10^6$
5(8)	32	258	3481	26	97356959	$2.627*10^9$
6(7)	32	241	3481	26	122103423	$2.534*10^9$

Table 26.3. The results for approximate algorithms

P(K)	N	L	N_{clk}^{h}	N_{clk}^{s}	K_a	R
1(4)	32	53	24770	$0.41*10^6$	4	
2(3)	18	21	4885	$0.12*10^6$	3	
3(4)	28	50	17612	$0.31*10^6$	4	
4(3)	18	12	3861	$0.067*10^6$	3	1
5(8)	32	258	20388	$0.506*10^6$	10	
6(7)	32	241	21578	$0.53*10^6$	10	
5(8)	32	258	25877	$1.04*10^6$	9	2
6(7)	32	241	24803	$0.625*10^6$	9	
5(8)	32	258	2680182	$1.11*10^9$	8	3
6(7)	32	241	3318341	$72.67*10^6$	8	
6(7)	32	241	106231355	$2.24*10^9$	7	4

The first problem (1) is the colouring of central and western European map. The next three problems (2–4) are encodings of micro-operations in microinstructions. The last two examples (5,6) were generated randomly.

We can summarise the results of experiments and analysis of alternative realisations as follows:

1. In average, the FPGA-based implementations require one order of magnitude less clock cycles comparing to software implementations.
2. Approximate algorithms make possible to obtain either optimal or near-optimal results after just a few iterations. Thus, they are very well-suited for many practical applications. Besides, they enable us to reduce very significantly the execution time.
3. The analysis of execution time has shown that approximate algorithms are very well suited for run-time solution of the graph colouring problem.
4. Preliminary tests of the considered algorithms in hardware description language based projects (VHDL, in particular) show that the number of FPGA slices can be significantly reduced.
5. The considered projects are easily scalable making possible to customise the required complexity and to link the projects with embedded systems for robotics working in real-time environment.
6. The used architecture (see Fig. 26.2) can easily be retargeted to solve some other combinatorial problems, such as matrix covering or the Boolean satisfiability [8].

26.7 Conclusion

The first important contribution of the chapter is the presentation of the known graph colouring problem in such a way that makes possible easy implementation of projects for solving this problem in commercially available FPGAs. The second contribution is the analysis which permits to conclude that FPGA-based projects are more advantageous than relevant software projects executing on general-purpose computers (in terms of the number of clock cycles). The remaining significant issues considered in the chapter are the exact algorithm for graph colouring and the derived approximate algorithms, which can be used for solving numerous engineering problems in the scope of robotics and embedded systems.

References

1. Celoxica products (visited on 03/04/2007). [Online] http://www.celoxica.com
2. Culberson J (visited on 03/04/2007) Graph coloring page. [Online] http://www.cs.ualberta.ca/~joe/Coloring/index.html
3. Ezick J (visited on 03/04/2007) Robotics. [Online] http://dimacs.rutgers.edu/REU/1996/ezick.html
4. Goossens G, Van Praet J, Lanneer D, Geurts W, Kifli A, Liem C, Paulin PG (1997) Embedded software in real-time signal processing systems: design technologies. Proceedings of the IEEE, 85(3):436–454
5. Lee TK, Leong PHW, Lee KH, Chan KT, Hui SK, Yeung HK, Lo MF, Lee JHM (1998) An FPGA implementation of GENNET for solving graph coloring problems. In: Proc. IEEE Symposium on Field-Programmable Custom Computing Machines-FCCM, USA, pp 284–285
6. Pochet LM, Linderman M, Kohler R, Drager S (visited on 03/04/2007) An FPGA based graph coloring accelerator. [Online] http://klabs.org/richcontent/MAPLDCon00/Papers/Session_A/A2_Pochet_P.pdf
7. Rosen KH (ed) (2000) Handbook of Discrete and Combinatorial Mathematics. CRC Press, USA
8. Skliarova I (2004) Reconfigurable Architectures for Combinatorial Optimization Problems. Ph.D. thesis, University of Aveiro, Portugal
9. Sklyarov V (2004) FPGA-based implementation of recursive algorithms. Microprocessors and Microsystems, Special Issue on FPGAs: Applications and Designs, 28(5–6):197–211
10. Sklyarov V, Skliarova I, Pimentel B (2006) Modeling and FPGA-based implementation of graph coloring algorithms. In: Proc. 3rd International Conference on Autonomous Robots and Agents-ICARA'2006, Palmerston North, New Zealand, pp 443–448
11. Subramonian V, Huang HM, Xing G, Gill C, Lu C, Cytron R (visited on 03/04/2007) Middleware specialization for memory-constrained networked embedded systems. [Online] http://www.cs.wustl.edu/~venkita/publications/rtsj_norb.pdf
12. Wu YL, Marek-Sadowska M (1993) Graph based analysis of FPGA routing. In: Proc. European Design Automation Conference, Germany, pp 104–109
13. Xilinx products (visited on 03/04/2007). [Online] http://www.xilinx.com
14. Zakrevskij AD (1981) Logical Synthesis of Cascade Networks. Science, Moscow

27

Balancing Sociality in Meta-agent Approach

Oomiya Kenta[1], Miyanishi Keiji[2], Suzuki Keiji[1,3]

[1] Institute of System Information Science, Future-University Hakodate 116-2, Kamedanakano, Hokkaido, Japan
[2] Information Science Research Center
[3] Core Research for Evolutional Science and Technology (CREST, JST)
g3107001@fun.ac.jp, miyanishi@isrc.co.jp, suzukj@fun.ac.jp

Summary. This paper shows an agent based approach to solve the Tragedy of the Commons. This game is one of the well-known problems that involve sharing limited common resources. In this paper, to control the usage of common resources, we employ Levy Based Control Strategy and extend it with Autonomous Role Selection. In this approach, what levy plan of meta-agents should be used is very important to avoid the tragedy situation. Accordingly, we apply Genetic Algorithm (GA) to each agent to obtain evolutionally suitable levy plan. As a result, we show effectiveness of this approach and it depends on a fitness function of GA. Therefore, we examine the various fitness functions considered for a social balance and the importance of agents' role by simulations.

Keywords: Meta-agent, The Tragedy of the Commons, levy-based control.

27.1 Introduction

The Tragedy of the Commons (TC) [2] is a well-known game problem, which is one of the N-person social dilemmas [4]. The characteristic of this game is known that no technical solution exists. In this paper, we extend this game with Levy Based Control Strategy by a meta-agent (LBCS) [3, 5] and Autonomous Role Selection (ARS). This strategy is a kind of approaches changing the problem structure. They can avoid the tragedy situations because meta-agents, which are treated as local government, can change the problem structure by charging levies on activities of agents. The issues in this approach are i) how meta-agents are introduced and ii) what levies should be used. In this paper, ARS is introduced to the strategy so as to solve above i). This allows agents to become meta-agents autonomously. To solve above ii), we apply Genetic Algorithm (GA) to each agent so as to acquire suitable levies evolutionally. We must consider what fitness function of GA is better. In this paper, we employ the fitness function which evaluates the sociality of the agents and the importance of roles. This function can evaluate various ratio of above two elements by configure the parameters of the fitness function. That is, we can try various fitness functions so as to examine the effective area of this approach.

O. Kenta et al.: *Balancing Sociality in Meta-agent Approach*, Studies in Computational Intelligence (SCI) **76**, 233–239 (2007)
www.springerlink.com

27.2 The Tragedy of the Commons

The Tragedy of the Commons is a well-known game problem, which is one of the N-person social dilemmas [4]. This game enables us to analyse the behaviour of players who share common limited resources. As common resources are limited, the higher activities of agents to gain higher payoffs will result in lower payoffs. Because all agents have individual rationality, they invariably select the higher activities. That is, they can not avoid the tragedy situation.

Here, we show an explanation of this game. There are N agents. They can get payoffs by consuming limited common resources. Their consumption level is represented by $A = \{a_0, a_1, \ldots, a_n\}$. The activity of the agent i, a^i, is to select consumption level. One of the general forms of the payoff function in TC is as follows [1].

$$Payoff(a^i, TA) = a^i(|A| \times N - TA) - 2a^i, \ where\ TA = \sum_{j=1, i \neq j}^{N} a^j \tag{27.1}$$

This payoff function represents a payoff which an agent i can get. In this equation, TA (Total Activity) indicates a quantity of total consumption; |A| is the number of consumption level. Important characteristics of this payoff function are i) payoff values can not be decided by one agent, ii) $Payoff(x, TA)$ are always smaller than $Payoff(x + 1, TA)$. These characteristic cause a social dilemma.

The characteristic of this game is known that no technical solution exists. Therefore, to solve this game, players should change individual rationality to other types of rationality or change the problem structure by changing the payoff function. LBCS is a kind of an approach changing the problem structure. In this paper, we propose an approach which extends LBCS with ARS.

27.3 The Extended Problem

In this paper, we extend LBCS with ARS. This section shows the details of our proposed approach.

27.3.1 The Levy-based Control Strategy by a Meta-agent

This approach is a kind of multi-agent approach. This approach can solve the problem by changing problem structures. A Meta-agent is introduced into the game as an agent which charges levy on other agents, which are called Players. The purpose of the meta-agent is to maximise personal revenue from the agents. That is, the meta-agents obey individual rationality. The meta-agent has a levy plan (27.2) which consists of levy values $\{Lv_0, Lv_1, \ldots, Lv_n\}$ corresponding to consumption level $A = \{a_0, a_1, \ldots, a_n\}$. The limits of the levy values are payoff values. The levy plan of the meta-agent k is represented as follows:

$$LP^k = \left\{Lv_j^k \,|0 \le j \le |A|\,\right\} \qquad where\ 0 \le Lv_j^k \le Payoff(a_j, TA) \tag{27.2}$$

Charging levies on players changes the definition of gains of players. In our approach, gains of players are called rewards. Gain of the meta-agent is called revenue. The revenue is sum of levies from players. The rewards is remainder which subtracts the levy corresponding to the activity from the payoff.

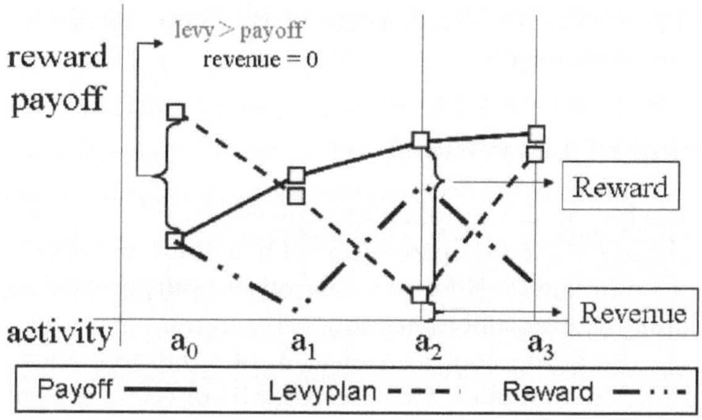

Fig. 27.1. Relation between levies and the payoff function

Figure 27.1 shows the relation between a levy plan and the payoff function. The horizontal axis indicates the activities and the vertical axis indicates gain value. A solid line indicates payoff values. A dash line indicates a levy plan. A dashed-two dotted line indicates rewards for agents. In this figure, the second highest activity a_2 produce the highest gain. If the agents obey individual rationality, they select activity a_2. Because suitable levy plans can prevent selecting the highest activity a_3, the levy based control strategy can avoid the tragedy situations.

When we consider introducing this strategy, there are two issues, i) how to introduce the meta-agent and ii) how to set a suitable levy plan. The first issue is described in the next subsection. The second issue is described in the subsection 27.3.3.

27.3.2 Autonomous Role Selection

In this paper, ARS is introduced into LBCS so as to solve the above problem i). That is, multiple meta-agents are allowed. Therefore, the definition of the agents in TC is changed. In our proposed approach, agents can select personal role, player or meta-agent, so as to maximize personal gains. The activities of extended agents are defined as follows.

$$a^i \in \{A_{player}, A_{meta}\} \text{ Wherer } A_{player} = \{a_j \,|0 \le j \le |A|\}, \quad A_{meta} = \{a_{meta}\} \tag{27.3}$$

When multiple meta-agents propose levy plans, a social levy plan (SLP) is generated from proposed levy plans. SLP consists of minimum levies on each activity. The meta-agents who propose the minimum levies can obtain revenues. SLP is represented as follows.

$$SLP = \left\{Lv_j^{\min} \,|0 \le j \le |A|\right\} \qquad where\ Lv_j^{\min} = \min\left\{Lv_j^1, \ldots, Lv_j^N\right\} \tag{27.4}$$

The definitions of rewards and revenues are changed by introducing SLP. These are represented as follows. $N(a_j)$ means the number of agents who select activity a_j.

$$Revenue^k = \begin{cases} \sum_{j=0}^{|A|} N(a_j) \cdot Lv_j^{\min} & (if Lv_j^{\min} = Lv_j^i) \\ 0 & (if Lv_j^{\min} \ne Lv_j^i) \end{cases} \tag{27.5}$$

$$Reward^i = Payoff(a^i, TA) - Lv_{a^i}^{\min} \tag{27.6}$$

The SLP can moderate selfish charging of the meta-agents by introducing the principle of market mechanism.

27.3.3 Evolution of Levy Plans

The property of the levy plan is important. Apparently, it is possible to fix a suitable levy plan by analysing the game properties. Then the levy plan can be embedded in the simulation beforehand. However, one of the purposes of agent-based simulations is to examine the autonomous property of the agents in a social environment. Accordingly, the autonomous acquisition of a suitable levy plan is desirable. To search for a suitable levy plan, Genetic Algorithm (GA) is applied.

In this paper, GA is applied to each agent. That is, each agent has N_{LP} levy plans which evolve independently. Gene type of these levy plans consists of levies on each activity. The meta-agent selects the highest evaluated gene as its levy plan. The fitness function of the levy plans is as follows.

$$E_i = \left(\frac{\alpha \cdot Revenue_i + \beta \cdot \mathrm{Reward}_i}{Reward_{i-worst}}\right)^a + \left(\frac{\sum_{j=1}^{N} Reward_j}{\sum_{j=1}^{N} Reward_{j-worst}}\right)^b \tag{27.7}$$

The first term evaluates selfish behaviours of agents. The second term evaluates altruistic behaviours of agents. $\mathrm{Reward}_{\mathrm{worst}}$ means the lowest gain in the tragedy situation. Parameter-Alpha is a weighting factor of the revenue. Parameter -Beta is a weighting factor of the reward. Parameter-A is a weighting factor of the first term and parameter-B is a weighting factor of the second term. Ratio between parameter-A and B means the importance of agent's role. Ratio between parameter-alpha and beta means ratio of sociality. To configure these four parameters, we can search how fitness function is better for this simulation. The details of settings about parameters are described in Sect. 27.4.

27.3.4 The Proposed Simulation

Our proposed simulation consists of two steps, Generation of levy plans and Trial Step. In the trial step, the extended game is repeated many times so as to evaluate every levy plan.

In the first step, levy plans of each agent are generated with GA. That is, each agent has levy plans and uses the levy plan of those.

In the trial step, each levy plans must be evaluated so as to apply GA. That is, the game is repeated $N_{LP} \times N$ times in order to evaluate every levy plans. Firstly, levy plans which other agents have are randomly selected. Secondly, the agents decide personal role which maximize personal gains in turn. Thirdly, the rewards and revenues are calculated. Fourth, the evaluation of the evaluating levy plan is calculated.

27.4 Simulation and Results

In this section, in order to examine the effectiveness of the proposed methods the game simulations are executed.

27.4.1 The Settings of Simulation

In this subsection, settings of this simulation are showed. The payoff function is set as follows:

$$Payoff(a^i, TA) = a^i(|A| \times N' - TA) - 2a^i, where\ TA = \sum\nolimits_{j=1, i \neq j}^{N} a^j \qquad (27.8)$$

where $N\prime$ is the number of agents, excluding the meta-agents.

The parameters in the simulation are as follows: the number of agents, N, is 12. The activity of the agent is $a_j \in \{a_0 = 0,\ a_1 = 1, \ldots,\ a_5 = 5,\ a_{\mathrm{meta}} = -1\}$. The initial activity is $a_1 = 1$. Each agent has 30 chromosomes. The length of the revising step is 12. To evaluate a chromosome, the game is repeated four times using the same chromosome set. Furthermore, it is repeated three times using other chromosome sets of other agents. That is, the game is iterated 12 times to evaluate one levy plan. Crossover and mutation are applied to the chromosomes. The crossover rate is 10% and the mutation rate is 5%. Under these parameters, the evolution of the levy plans of the agents proceeds until 50 generations.

In this paper, to confirm the effectiveness of the extended evaluation, the simulation is iterated 81(= 9 times 9) times using sets of α, β, a and b. The details of parameter sets are indicated in Figure 27.2 and Figure 27.3.

27.4.2 Results of the Simulation

Figure 27.2 is the map of averages of rewards in the various parameter sets. Figure 27.3 is the map of averages of the activities. These figures are the results of conclusive results of the proposed simulation. The vertical axis represents the ratio of sociality. High selfishness is represented by low values on the vertical axis. High altruism is represented by higher values on the vertical axis. The horizontal axis represents the importance of the roles of the agents. The left side of the horizontal axis represents high importance of the role of meta-agents. The right side of the horizontal axis represents high importance of the role of the players. In these figures,

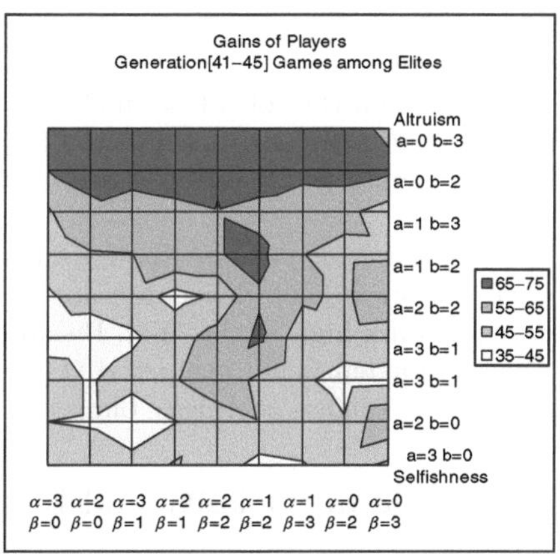

Fig. 27.2. Map of players' reward

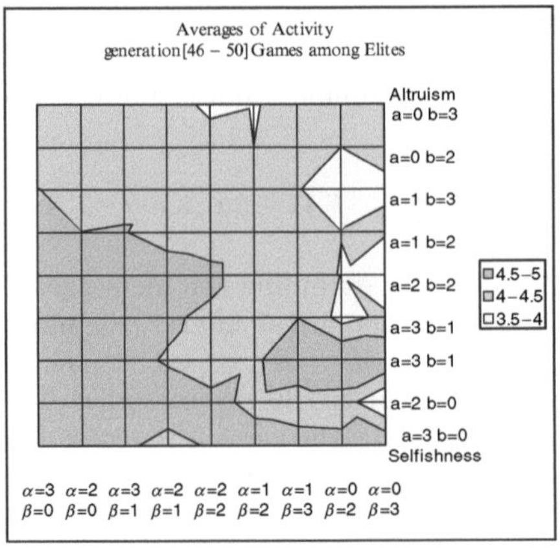

Fig. 27.3. Map of average of players' activities

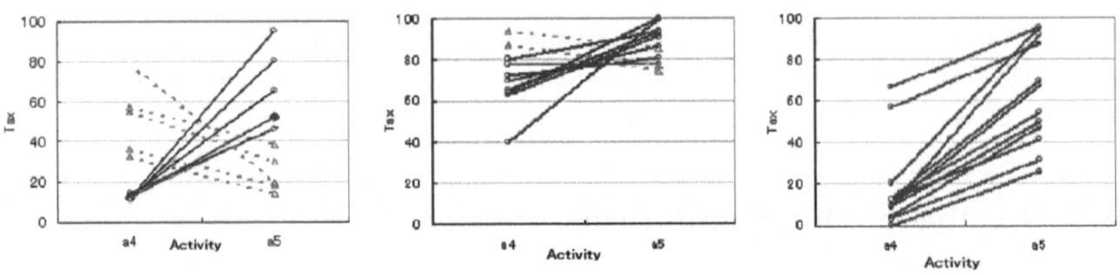

Fig. 27.4. An example of acquired levy plans

deep colours represent high values. In Figure 27.2, the worst gain 35–45 represents the tragedy situation. Therefore, the effectiveness of the meta-agents shows in areas excluded the worst area.

27.4.3 Features in Evaluation of the Role of the Agents

In the horizontal axis of Figures 27.2 and Figure 27.3, the colour of the right side is deeper than the colour of the left side. That is, these figures show that the role of the player is more important than the role of the meta-agent. This case improves not only the rewards but also the revenues, because the levy plans which can not moderate the activities of the agents stably exist. Figure 27.4(a) shows that such a few levy plans exist.

The area of ($\alpha = 0,\ \beta = 2$) is an exception. Although the activities of the players are moderated in this area, the rewards of players are not high. On the other hand, the revenues of the meta-agents are high in the area. Because the parameter α is 0 in the area, the role of the meta-agent is not evaluated. Therefore, the levies do not become low values. Figure 27.4(b) shows such levy plans.

27.4.4 Features in the Ratio of Sociality and Individuality

In the vertical axis, the rewards of the players are high in the area of ($a = 0,\ b = 2$). This parameter set shows that evaluation of the levy plan does not depend on the first term in the fitness function, which represents the selfish behaviour. Therefore, in this case, the levy plan is evaluated by the rewards of all players including own reward. Therefore, levy plans which can not improve total gains of the agents are evolutionarily eliminated. That improves the rewards of the players. By examining the levy plans which appear in these parameter settings, we can confirm the levy plans which can moderate the highest activities of the agents, in this case $a_5 = 5$. Figure 27.4(c) shows such levy plans.

27.5 Conclusions

This work addresses the game of the Tragedy of the Common by a social agent-based approach with the Levy Based Control Strategy. This paper evaluates the importance of the role of the agents and the ratio of sociality in the coevolution. Throughout the experiment, we show that the proposed approach performed well and could avoid the tragedy situation. Furthermore, we indicated the optimum values of the parameters in the evaluation of the GA. The features of these parameters emphasise the role of players and altruism. The characteristics of the acquired levy plans must be analysed. In particular, the relationship with the Nash equilibrium should be clarified. These points will be considered in a future work.

27.6 Acknowledgement

This work is supported by Core Research for Evolutional Science and Technology, Japan Science and Technology Agency (CREST, JST).

References

1. Keiji Suzuki, "Dynamics of Autonomous Changing Roles in social Dilemma Games", *The IJCIA-03 Workshop on Multiagent for Mass User Support*, pp 19–25 (2003).
2. G. Hardin, "The tragedy of the commons", *Science (162)*, pp 1234–1248 (1968).
3. T. Yamashita, K. Suzuki and A. Ohuchi, "Distributed Social Dilemma with Competitive Meta-players", *Int. Trans, in Operational Research*, 8(1), pp 75–88 (2001).
4. X. Yao, "Evolutionary stability in the N-person prisoner's dilemma", *Biosystems*, 37, pp 189–197 (1996).
5. Tomohisa Yamashita, Keiji Suzuki and Azuma Ohuchi, "Dynamics of Social Behavior with Gini Coefficient in Multiple Lake Game with Local Governments", Proceedings of IEEE International Conference on Systems, Man, and Cybernetics (SMC'99), Tokyo (Japan), pp. 586–591 (1999)

28

Complex Stability Radius for Automatic Design of Gain-Scheduled Control Part I: Theoretical Development

Rini Akmeliawati[1] and Iven M.Y. Mareels[2]

[1] Monash University Malaysia, School of Engineering, No. 2 Jalan Universiti Bandar Sunway, Selangor, Malaysia rini.akmeliawati@eng.monash.edu.my
[2] University of Melbourne, Dept. Electrical and Electronic Engineering, Parkville, VIC, Australia i.mareels@unimelb.edu.au

Summary. In this chapter, we present a systematic and automatic gain-scheduled control design technique using complex stability radius. The technique provides automatic selection of sets of operating points for a gain-scheduled control.

Keywords: Gain-scheduled control, automation, complex stability radius, stability robustness.

28.1 Introduction

Gain-scheduled control is one of the most common control techniques used in practice. This control technique is based on linear control theory but results in a non-linear control. In general, gain-scheduling control is obtained points as follows: (1) selecting a set of operating points, (2) designing (linear) controller for each operating point selected, (3) scheduling or interpolating the resulting family of the controllers to obtain a global controller.

This technique is initially known as an ad-hoc technique, whereby design validation and verification relies much on extensive simulations. Not until the last decade that analytical and mathematical justifications are available to demonstrate the stability and performance robustness of GS control [1, 2].

Many studies can be found on various types of gain-scheduled control, such as switching control by control Lyapunov function [3], linear-fractional transformation (LFT) [4, 5], linear-parameter varying (LPV) control [6, 7]. Most of those are summarised in [8, 9]. All those techniques are based on assumption that the set of operating points are known apriori. The issues on how many operating points to use and how to obtain the *appropriate* set of operating are overlooked and have never been addressed. Our proposed technique will address these issues. In practice, this first step of GS control is still done based on heuristics and extensive simulations. Prior

R. Akmeliawati and I.M.Y. Mareels: *Complex Stability Radius for Automatic Design of Gain-Scheduled Control Part I: Theoretical Development*, Studies in Computational Intelligence (SCI) **76**, 241–249 (2007)
www.springerlink.com

knowledge of the system to be controlled becomes necessary to select the operating points. Thus, GS control results in a time-consuming and expensive technique.

In [10] an automation procedure for a missile autopilot using the structured singular value μ is proposed. The results indicate conservatism due to the inadequacy of the μ-analysis in analysing mixed uncertainties (parametric and dynamic) of the missile dynamics [10].

In this chapter a systematic way of selecting the operating points for GS control system with guaranteed stability and performance robustness is presented. With the proposed automatic gain-scheduled control, we construct the grid points at the scheduling-variable space based on the complex stability radius. Thus, for any trajectories, the closed-loop system will guarantee the stability and performance robustness. This technique is an extension to the earlier work in [1]. The complete process from selection of operating points that guarantees stability and performance robustness to design and verification is addressed.

Our proposed automatic GS technique is similar to the stability-based switching controller developed in [3]. The difference is that our proposed technique explores the idea of complex stability radius whereas in [3] the technique uses equilibria-dependent Lyapunov functions. The advantage of using the complex stability radius is that it can handle a wider class of time variations [14]. This is not the case with the equilibria-dependent Lyapunov functions. Furthermore, in our proposed technique, there is no need to compute the Lyapunov function for each equilibrium or operating points although the stability robustness is achieved in Lyapunov sense. Our technique also addresses the problem encountered in [10] whereby μ-analysis cannot handle mixed parametric and dynamic uncertainties as mentioned earlier.

This chapter contains the theoretical development of the proposed technique. It is the extended version of the earlier paper [2]. The structure of this chapter is as follows. In Sect. 28.2 we will briefly introduce mathematical notations and definitions. The definition of complex stability radius is presented in Sect. 28.3. Section 28.4 presents the automatic GS algorithm. Finally we summarise this chapter in Sect. 28.5. Example on how the proposed method is implemented on a dynamical system is provided in Part II.

28.2 Mathematical Preliminaries

Here we introduce some basic mathematical terminology and definitions that are used in the rest of this chapter.

1. $\mathbb{F} = \mathbb{R}$ or $\mathbb{C}$. $\mathbb{R}$: the set of real numbers. $\mathbb{C}$: complex plane.
2. Let S be a set. For $S \subset \mathbb{R}^n$, ∂S , and S^o (or $int(S)$) are denoted as the *boundary* and the *interior* of S, respectively. $\bar{S}$ is the complement of S.
3. Given a family of sets ${S_j}_{j \in J}$, $S \subset X$ the *intersection* of this family is defined as a set $\cap_{j\in J} S_j = \{x \in S_j, \forall j \in J\}$. The *union* of this family is defined as a set $\cup_{j \in J} S_j = \{x \in S_j,\ j \in J\}$.
4. Let $T : D \to W$. The set $Im(T) := \{T(x) \in W; x \in D\}$ defines the *image* of T.
5. $\mathbb{C}_- = \{(x, y) \in \mathbb{R}^2; x < 0\}$; $\mathbb{C}_- = \{z \in \mathbb{C}|Real(z) < 0\}$.
6. $\mathbb{C}_+ = \mathbb{C}\backslash\mathbb{C}_-$, is the closed right half plane.
7. $L^p(I; \mathbb{F}^{n\times m})$ = the set of $p-$integrable functions $f : I \to \mathbb{F}^{n\times m}, I \subset \mathbb{R}$ an interval, $p \geq 1$.

8. $L^\infty(I;\mathbb{F}^{n\times m})$ = set of functions $f: I \to \mathbb{F}^{n\times m}$ that are essentially bounded on the interval $I \subset \mathbb{R}$.
9. $L^{pb}(I;\mathbb{F}^{n\times m})$ = set of piecewise continuous and bounded maps $M(.) : I \to \mathbb{F}^{n\times m}, I \subset \mathbb{R}$ an interval.
10. $\|f\|_{L^\infty(I;\mathbb{F}^{n\times m})} = \sup_{t\in I}\|f(t)\|$.

Consider a non-linear system of the form

$$\dot{x}(t) = f(x(t), u(t)), \qquad x(0) = x_o, \quad t \in \mathbb{R}, \tag{28.1}$$

where $x(t) \in \mathcal{D} \subseteq \mathbb{R}^n$ is the system state vector, $\mathcal{D}$ is an open set, $u(t) \in \mathcal{U} \subseteq \mathbb{R}^m$ is the control input, $\mathcal{U}$ is the set of all admissible controls such that $u(.)$ is a measurable function with $0 \in \mathcal{U}$, and $f : \mathcal{D} \times \mathcal{U} \to \mathbb{R}^n$ is continuously differentiable on $\mathcal{D} \times \mathcal{U}$.

Definition 1. *$x^* \in \mathcal{D}$ is an* equilibrium point *of (28.1) if there exists $u^* \in \mathcal{U}$ such that $f(x^*, u^*) = 0$.*

Definition 2. *The trajectory $x(t) \in \mathcal{D}$ denotes the solution of (28.1) with $x(0) = x_0$ evaluated at time t.*

28.3 Complex Stability Radius

In this section, we summarise the basic and general concepts on the stability radius. Consider a linear (time-varying) system

$$\dot{x}(t) = A(t)x(t), \quad t \geq 0, \tag{28.2}$$

where $A(.) \in L^{pb}(\mathbb{R}_{\geq 0};\mathbb{F}^{n\times n})$ is stable, piecewise continuous and bounded.

Stability radius is a measure of stability bound of a system. The stability radius for time-invariant systems was introduced by Hinrichsen and Pritchard [15]. The concept is extended for time-varying systems by Hinrichsen et. al [16] and for time-varying systems with multiperturbations by Hinrichsen and Prichard in [17].

We consider additive non linear perturbations. Thus, the perturbed system is of the form:

$$\dot{x}(t) = A(t)x(t) + \mathcal{E}(t)\mathcal{D}(\mathcal{H}(.)x(.))(t), \tag{28.3}$$

where the perturbation structure is represented by $\mathcal{E}(.) \in L^{pb}(\mathbb{R}_{\geq 0};\mathbb{F}^{n\times l})$, $\mathcal{H}(.) \in L^{pb}(\mathbb{R}_{\geq 0};\mathbb{F}^{k\times n})$ and the perturbation operator $\mathcal{D} : L^2(\mathbb{R}_{\geq 0},\mathbb{F}^p) \to L^2(\mathbb{R}_{\geq 0},\mathbb{F}^l)$.

Three different perturbation classes; time-varying linear perturbation $\mathcal{P}_{lt}(\mathbb{F})$, time-varying non-linear perturbation $\mathcal{P}_{nt}(\mathbb{F})$ and dynamical perturbation $\mathcal{P}_{dyn}(\mathbb{F})$ as defined in detailed in [14] are applicable in our context. This is because the complex stability radius r_C is essentially the sharp bound for the norm of the perturbation so that the global L^2 stability of the perturbed system (28.3) is preserved as long as $||\mathcal{D}|| < r_C$.

The nominal system (28.2) is subjected to time-varying perturbations of the form $A(.) \mapsto A(.) + \mathcal{E}(.)\Delta(.)\mathcal{H}(.)$, where $\Delta(.) \in L^\infty(\mathbb{R}_{\geq 0};\mathbb{F}^{l\times k})$. The resulting perturbed system is

$$\dot{x}(t) = (A(t) + \mathcal{E}(t)\Delta(t)\mathcal{H}(t))x(t). \tag{28.4}$$

$A(t)$ may be interpreted as $A_{cl}(t)$, a closed-loop system matrix, where $(A(t), B(t))$ is assumed to be stabilisable, $A_{cl}(t) = A(t) - B(t)K(t)$, $K(t) \in \mathbb{F}^{m \times n}$ is a stabilising controller. $\sigma(A_{cl}(t)) \subset \mathbb{C}_-$. In this case the perturbation Δ contains plant and input signal perturbations.

Definition 3. *[15] The stability radius of a 'frozen' A_{cl} is defined by*

$$r_F(A_{cl}; \mathcal{E}, \mathcal{H}, \mathbb{C}_+) = \inf\{\|\triangle\|; \triangle \in \mathbb{F}^{l \times k}, \sigma(A_{cl} + \mathcal{E}\triangle\mathcal{H}) \cap \mathbb{C}_+ \neq \emptyset\}.$$

The complex stability radius $r_c(A; \mathcal{E}, \mathcal{H})$ of the system (28.2) with respect to the perturbation structure $(\mathcal{E}, \mathcal{H})$ is the spectral norm $||\Delta||$ of a smallest destabilising perturbation matrix $\Delta \in \mathbb{C}^{l \times k}$. Complex stability radius measures the robustness of stable linear systems under complex (parameter) perturbations. Obviously, complex stability radius is more conservative than its real counterpart as the later is a measure of robustness under real perturbations only.

The main result in [14] shows that the structured stability radius of the time-varying system is close to the infimum of the structured stability radii of all 'frozen' systems; or precisely, for fixed $\tau \geq 0$, with $(A_{cl}(\tau), \mathcal{H}(\tau))$ observable, then there exists $\delta > 0$ and $\rho > 0$ such that $\|\dot{A}_{cl}(\tau)\| + \|\dot{\mathcal{E}}(\tau)\| + \|\dot{\mathcal{H}}(\tau)\| < \delta$ and $\rho \leq r_{c,dyn}(A_{cl}; \mathcal{E}(.), \mathcal{H}(.))$.

Theorem 4.3 in [14] proves that for a system (28.2) and $(\mathcal{E}(.), \mathcal{H}(.))$ defining the three classes of perturbations, for $\mathbb{F} = \mathbb{R}$ or $\mathbb{F} = \mathbb{C}$, the followings are true.

$$r_{\mathbb{F}} \geq r_{\mathbb{F},lt} \geq r_{\mathbb{F},nt} \geq r_{\mathbb{F},dyn}, \tag{28.5}$$

and if A or A_{cl}, $\mathcal{E}(.)$ and $\mathcal{H}(.)$ are constant matrices then

$$r_{C}(A; \mathcal{E}, \mathcal{H}) = r_{C,dyn}(A; \mathcal{E}, \mathcal{H}) = \min_{\omega \in \mathbb{R}} ||\mathcal{H}(sI - A)^{-1}\mathcal{E}||^{-1}. \tag{28.6}$$

In case a closed-loop system A_{cl} is used instead of A in equation (28.6), where A_{cl} is exponentially stable.

Our proposed automation exploits the complex stability radius to determine which operating points should be selected for the controller(s) design. The result of this automation is a family of operating points with associated controllers such that the stability and performance robustness of the overall design is guaranteed.

28.4 Automation of Gain-Scheduled Control

We describe our gain-scheduled control as follows. Let $\Omega \subset \mathbb{R}^q$, a compact set, denotes the set of GS parameters so that for every $\theta \in \Omega$ we have an equilibrium (x_θ, u_θ) of (28.1), i.e. , $f(x_\theta, u_\theta) = 0$. More precisely, we assume that

$$\begin{aligned} s \times c : \Omega &\rightarrow \{(\bar{x}, \bar{u}) \in \mathbb{R}^n \times \mathbb{R}^m | f(\bar{x}, \bar{u}) = 0\}, \\ \theta &\mapsto (s(\theta), c(\theta)) = (x_\theta, u_\theta). \end{aligned} \tag{28.7}$$

Let θ_0, $\theta_1 \in int(\Omega)$. Let $\theta_R(t)$ be a piecewise smooth, time-parameterised path contained in $int(\Omega)$ connecting $\theta_0 = \theta_R(0)$ and $\theta_1 = \theta_R(T)$, or

[3] $(A(t), B(t))$ is assumed to be stabilisable

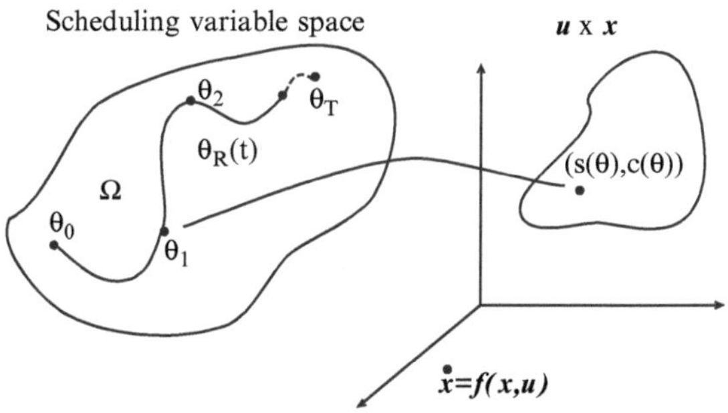

Fig. 28.1. Mapping from scheduling-variable space to $(x \times u)$ space

$\theta_R\,(.)\ :\ [0,T]\ \rightarrow\ \Omega, T\ >\ 0$. See Fig. 28.1. The system (28.1) is linearised about an equilibrium (operating point) (x_θ, u_θ) of (28.1) for each scheduling parameter $\theta \in \Omega$. Consider now the control law for the system (28.1)

$$u(t) = u_{\theta(t)} - K_\theta(x - x_{\theta(t)}), \tag{28.8}$$

where $K_{\theta_R} \in \mathbb{R}^{m\times n}$ is a feedback gain that strictly stabilises the linearised plant at the equilibrium $(x_{\theta(t)}, u_{\theta(t)})$. Hence, for each equilibrium (x_θ, u_θ) we use Taylor's Theorem to derive

$$f(x,u) = A_\theta[x - x_\theta] + B_\theta[u - u_\theta] + R(x_\theta, u_\theta, x, u),$$

where $A_\theta \in \mathbb{R}^{n\times n}$, $B_\theta \in \mathbb{R}^{n\times m}$, and $R : \mathbb{R}^{2n+2m} \rightarrow \mathbb{R}^n$ is a continuous function satisfying

$$\|R(x_\theta, u_\theta, x, u)\| \leq r_\theta(x,u) \left\| \begin{matrix} x - x_\theta \\ u - u_\theta \end{matrix} \right\|, \forall (x_\theta, u_\theta, x, u) \in \mathbb{R}^{2n+2m},$$

and $r_\theta(.,.) : \mathbb{R}^{n+m} \rightarrow \mathbb{R}_{\geq 0}$ is also continuous and $r_\theta(x,u) \rightarrow 0$ as $(x,u) \rightarrow 0$.

Based on the linear approximation of (28.1), i.e.

$$\frac{d}{dt}[x(t) - x_\theta] = A_\theta[x(t) - x_\theta] + B_\theta[u(t) - u_\theta], \tag{28.9}$$

a state feedback matrix $K_\theta \in \mathbb{R}^{m\times n}$ is designed so that (28.8) applied to (28.9) yields an exponentially stable closed loop system

$$\frac{d}{dt}[x(t) - x_\theta] = [A_\theta - B_\theta K_\theta][x(t) - x_\theta].$$

28.4.1 Automation

The following definition describes the automation process.

Definition 4. *[1] With reference to system (28.1), let $\theta \in \mathbb{R}^q$ be a scheduling variable, $\mathcal{P}$ be a partition of the operating envelope $\Omega \subset \mathbb{R}^q$ in scheduling-variable space and $\mathcal{U}_i$ is convex.*

$$\mathcal{P} = \{\mathcal{U}_i, i = 1 \cdots N\}; \mathcal{U}_i^\circ = \mathcal{U}_i \backslash \partial\mathcal{U}_i, \mathcal{U}_i^\circ \cap \mathcal{U}_j^\circ = \emptyset,\ i \neq j,\ \bigcup_i \mathcal{U}_i = \Omega.$$

Let θ_i be the center of $\mathcal{U}_i$, with associated controllers (K_i, u_i) and equilibrium x_i in the operating envelope, $u_i = c(\theta_i), x_i = s(\theta_i)$ (see (28.7) and (28.8)). The partition $\mathcal{P}$ is acceptable provided the induced GS controller is stabilising $\forall\theta(t)$ such that $\theta(t) \in \Omega$ and $\dot{\theta}(t)$ is sufficiently small $\forall t$, the system

$$\dot{x}(t) = f(x(t), u_{in(\theta(t))} + K_{in(\theta(t))}\hat{x}(t)), \tag{28.10}$$

$$\hat{x}(t) = x(t) - x_{in(\theta(t))}, \tag{28.11}$$

$$in(\theta(t)) = j \texttt{ if } \theta(t) \in \mathcal{U}_j, \tag{28.12}$$

is stable and achieves that $x(t)$ remains inside the operating envelope. Further, $\|x(t) - s(\theta(t))\| \ll 1$.

The partition $\mathcal{P}$ is computed by computing the center θ_i. Starting from a given θ_0 we compute the neighbouring center in an iterative manner, verifying that the new center is such that the changes are within the complex stability radius.

Proposition 1. *Using Definition 4, the automatic GS control of system (28.1) with the control law (28.8) proceeds as follows.*

1. *The first operating point and the corresponding scheduling variable(s) selection. (This operating point is generally the nominal point or a point where the non-linearities of the system are significant.)*
2. *Controller design. Feedback controller is designed for the linearised system to satisfy the performance and stability requirement.*
3. *Determine $\mathcal{E}$, $\mathcal{H}$ matrices of the system's perturbation and compute the complex stability radius, ρ_i.*
4. *Construction of stability region, i.e. a ball around θ_i contained in stability radius, more precisely (see Fig. 28.2):*
   a) $\mathcal{E}\Delta_\theta\mathcal{H} = (A_i - B_iK_i) - (A_{s(\theta)} - B_{s(\theta)}K_i)$,

   b) $r(\theta) = \|\Delta_\theta\|$,

   c) $\Theta_i = \{\theta : r(\theta) \leq \rho_i\}$,

   d) $B_{\theta_i}(\alpha) = \{\|\theta - \theta_i\| \leq \alpha\}$ (for a given norm) with polygonal level curve.

   e) Find maximal $\bar{\alpha}$ such that $B_{\theta_i}(\bar{\alpha}) \subset \Theta_i$.
5. *Choose θ_{i+1} on the boundary of $B_{\theta_i}(\bar{\alpha})$ and repeat the process.*
6. *Optimise the θ_i lattice to obtain minimal overlap, and a full partition.*

With the notion of complex stability radius in constructing the grid in the scheduling-variable space, exponential stability of the 'frozen' system guarantees exponential stability of the time-varying system and the non-linear system (28.1).

To support Proposition 1, we use Theorem 1.

Theorem 1. *Consider $\epsilon > 0$ and Lemma 1. If our system (28.1) and the trajectory given as $\theta_R(t)$, and*

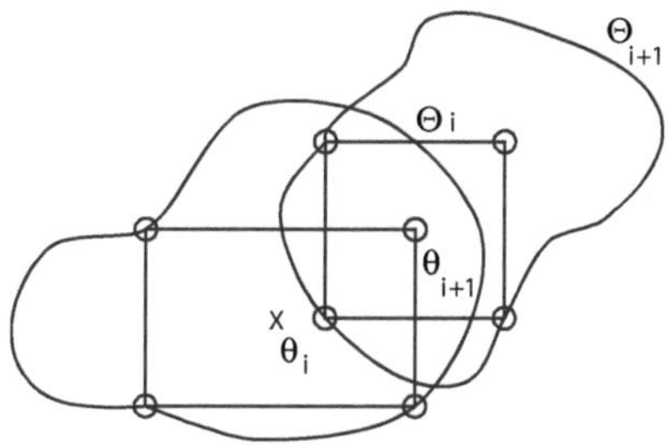

Fig. 28.2. Defining $B_{\theta_i}(\alpha)$ [1]

$$\left\| \frac{d}{dt}\theta_R(t) \right\| + \left\| \frac{d}{dt} x_{\theta_R(t)} \right\| < \delta_1, \quad \forall t \geq 0 \;\; and\, if \;\; \|e(0)\| < \delta_2,$$

for $\delta_1, \delta_2 > 0$ *sufficiently small, then the trajectory of the closed-loop system (28.15) satisfies*

$$\|x(t) - x_{\theta_R(t)}\| < \epsilon, \;\; \forall t \geq 0.$$

Proof: δ_1 sufficiently small guarantees that Lemma 1 holds and therefore (28.16) follows. Since Ω is compact and $\theta_R \mapsto K_{\theta_R}$ is continuous, there exists $k > 0$ such that

$$\|K_{\theta_R}\| \leq k \quad \forall \theta_R \in \Omega.$$

Substituting this equality and (28.13) into (28.16) yields, for all $t \in [0, \omega)$,

$$V(t, e(t)) \leq -[1 - 2c_2 k r_{\theta_R(t)}(e(t))]\|e(t)\|^2 + 2c_2\delta_1\|e(t)\|.$$

This proves that there is no finite escape time, i.e. $\omega = \infty$. For $\delta_2 > 0$ sufficiently small so that $r_{\theta_R(0)}(e(0)) < (4c_2k)^{-1}$, we may choose $\delta_1 > 0$ sufficiently small so that the claim follows. This completes the proof. ◇

Theorem 1 confirms that the scheduling variables should be slowly-varying and capture the plant's non-linearities. Thus, with our proposed automatic gain-scheduled control, we make use of the slow-varying 'restriction' of the scheduling-variable(s) and automatically select the operating points for designing the family of linear controllers $K(t)$ that provides (exponential) stability robustness to the closed-loop system regardless the types of perturbations and the trajectory of the system.

28.5 Conclusion and Future Work

The main contribution of the chapter is on addressing the issue of operating point selection in a systematic and automatic way with guaranteed stability robustness. The automation allows the GS control to be designed starting from the selection of operating points. The automation is based on the complex stability radius, which can handle a large class of perturbations or uncertainties, including non-linear, dynamics and time-varying. Thus, the stability robustness of the closed-loop system is guaranteed regardless of the types of trajectories (in the scheduling-variable space) that the closed-loop system is going through.

Future study is required to incorporate performance robustness measure directly to improve the presented result though this normally introduces more conservatism. Nonetheless, no matter how many operating points are used in the end, the automation proposed in this chapter can reduce the amount of time and cost in the design process of a gain-scheduled control.

28.6 Appendix

Lemma 1. *Consider the non-linear system (28.1). Ω and (28.7) are defined as before with $\theta(.) : \mathbb{R}_{\geq 0} \rightarrow \Omega, t \mapsto \theta(t)$ and (28.7) are continuously differentiable. For each $\theta_R \in im\ \theta_R(.)$, we set $A_{\theta_R} = \frac{\partial f(x_{\theta_R}, u_{\theta_R})}{\partial x}$, $B_{\theta_R} = \frac{\partial f(x_{\theta_R}, u_{\theta_R})}{\partial u}$. Let $K : Im\theta_R(.) \rightarrow \mathbb{R}^{n \times m}$, $\theta_R \rightarrow K_{\theta_R}$ and is continuously differentiable such that we have $\sigma(A_{\theta_R} - B_{\theta_R} K_{\theta_R}) \subset \mathbb{C}_{-}, \forall \theta_R \in Im\theta_R(.)$. Then, the followings are true:*

1. *$\|\frac{d}{dt}\theta_R(.)\|_{L^\infty(0,\infty)} < \epsilon$ for some $\epsilon > 0$ sufficiently small, then there exists a continuously differentiable map*

$$P_{\theta_R(.)} : \mathbb{R}_{\geq 0} \rightarrow \mathcal{H} = \{P \in \mathbb{R}^{n \times n} | P = P^T\}, t \mapsto P_{\theta_R(t)},$$

P_{θ_R} is well defined on Ω, and there exists constants $0 < c_1 < c_2$, such that

$$c_1 I_n \leq P_{\theta_R(t)} \leq c_2 I_n, \quad \forall t \geq 0, \tag{28.13}$$

and $P_{\theta_R(t)}$ satisfies

$$\frac{d}{dt} P_{\theta_R(t)} + \hat{A}^T_{\theta_R(t)} P_{\theta_R(t)} + P_{\theta_R(t)} \leq -I_n, \forall t \geq 0, \tag{28.14}$$

where $\hat{A}_{\theta_R(t)} := A_{\theta_R(t)} - B_{\theta_R(t)} K_{\theta_R(t)}$.

2. *Let $V(t,e) = e^T P_{\theta_R(t)} e$, along the closed-loop system*

$$\left.\begin{array}{c} \dot{x}(t) = f(x(t), u(t)) \\ u(t) = u_{\theta_R(t)} - K_{\theta_R(t)} e(t) \\ e(t) = x(t) - x_{\theta_R(t)} \end{array}\right\} \tag{28.15}$$

then,

$$\frac{d}{dt} V(t, e(t)) \leq -[1 - 2\|P_{\theta_R(t)}\| r_{\theta_R(t)} e(t) \|K_{\theta_R(t)}\|] \\ \|e(t)\|^2 + 2\|P_{\theta_R(t)}\| \|\frac{d}{dt} x_{\theta_R(t)}\| \|e(t)\|, \tag{28.16}$$

where

$$r_{\theta_R} = \sup_{0 \leq \lambda \leq 1} |f'(x_{\theta_R} + \lambda e, u_{\theta_R} + \lambda K_{\theta_R} e) - f'(x_{\theta_R}, u_{\theta_R})\|. \tag{28.17}$$

Proof: (1): When $t \mapsto \theta_R(t)$ is sufficiently slow, continuous differentiability of $f(.,.)$ and $(x.,u.)$, and compactness of Ω yield that $\frac{d}{dt}\hat{A}_{\theta_R(t)}$ is uniformly small. Now uniform boundedness of $\hat{A}_{\theta_R}$ in $\theta_R \in \Omega$ and exponential stability of $\hat{A}_{\theta_R}$ for every $t \geq 0$ yields, for $\|\frac{d}{dt}\hat{A}_{\theta_R(.)}\|_{L^\infty(0,\infty)}$ sufficiently small, uniform exponential stability of $\dot{\eta}(t) = \hat{A}_{\theta_R(t)}\eta(t)$. For a proof see e.g. Theorem 8.7 in [13]. Therefore, we may apply Theorem 8.7 in [13] to establish (28.13) and (28.14). This proves (1).
(2): By Taylor's Theorem, the error of the closed-loop system (28.15)

$$\dot{e}(t) = [A_{\theta_R(t)} - B_{\theta_R(t)}K_{\theta_R(t)}]e + \tilde{R}(K_{\theta_R(t)}, e) - \frac{d}{dt}x_{\theta_R(t)}, \qquad (28.18)$$

satisfies $\|\tilde{R}(K_{\theta_R(t)}, e)\| \leq r_{\theta_R}(e)\|K_{\theta_R(t)}\|\|e\|$ and by an estimate of the remainder of the Taylor expansion $r_{\theta_R}(e)$ is given as in (28.17). Now differentiation of $V(t, e(t))$ along (28.15) and invoking (28.18) and (28.15) yields (28.16). This proves (2). ◇

References

1. Shamma J, Athans M (1990) IEEE Trans. on Automatic control 35(8):898–907
2. Rugh W (1991) IEEE Control Systems Magazine 79–84
3. Leonessa A, et.al. (2001) IEEE Trans. on Automatic Control 46(1):17–28
4. Packard A (1994) Systems & Control Letters 22:79–92
5. Wu F, Dong K (2005) Proc. of the 2005 American Control Conference Vol. 4: 2851–2856
6. Scherer C (2001) Automatica 37(3):361–375
7. Mehrabian A R, Roshanian J (2006) Proc. of 1st Intl. Symp. on Systems and Control in Aerospace and Astronautics 144–149
8. Rugh W J, Shamma J S (2000) Automatica 36: 1401–1425
9. Leith D J, Leithead W E (2006) http://www.hamilton.ie /pubs_Leithead.htm visited on 1/9/2006.
10. Wise K, Eberhardt R (1992) Proc. of IEEE Conf. of Control on Applications 243–250
11. Akmeliawati R, Mareels I (1999) Proc. of the 14th IFAC World Congress 289–294
12. Akmeliawati R, Mareels I M Y (2006) Proc of the 3rd International Conference on Autonomous Robots and Agents (ICARA 2006) 665–670
13. Rugh W (1996) Linear System Theory. Prentice-Hall, New Jersey
14. Ilchmann A, Mareels I (2001) In: Adv. in Mathematical Systems Theory, A Vol. in Honor of Diederich Hinrichsen. Birkhauser Berlin
15. Hinrichsen D, Pritchard A J (1986) Systems & Control Letters 8:105–113
16. Hinrichsen D, Ilchmann A, Pritchard A J (1989) Journal of Differential Equations 82:219–250
17. Hinrichsen D, Pritchard A J (1991) Proc. European Control Conference Grenoble 1366–1371

29

Complex Stability Radius for Automatic Design of Gain-Scheduled Control Part II: Example on Autopilot Design

Rini Akmeliawati[1] and Iven M.Y. Mareels[2]

[1] Monash University Malaysia, School of Engineering, No. 2 Jalan Universiti Bandar Sunway, Selangor, Malaysia rini.akmeliawati@eng.monash.edu.my
[2] University of Melbourne, Dept. Electrical and Electronic Engineering Parkville, VIC, Australia i.mareels@unimelb.edu.au

Summary. In this chapter we demonstrate how our proposed automated gain-scheduling technique in Chapter 28 is implemented on an autopilot design for a civil aircraft model.

Keywords: Autopilot, gain-scheduled control, complex stability radius, automation.

29.1 Introduction

This chapter contains an elaborated example presented earlier in [2]. In this study we only consider the longitudinal dynamics of the aircraft. The automation is aimed to construct an acceptable partition in scheduling-variable space such that the gain-scheduled control guarantees stability and performance robustness.

This chapter is organised as follows. Section 29.2 describes the aircraft model in short and the control objectives. Section 29.3 presents the controller design followed by Sect. 29.4 whereby simulation results of closed-loop system are presented and discussed. Finally Sect. 29.5 provides the concluding remarks.

29.2 Aircraft Model and Control Objectives

The aircraft model used in this illustration is a twin-engine civil aircraft model that is developed by Group for Aeronautical Research and Technology in Europe (GARTEUR). The aircraft model is known as the research civil aircraft model or RCAM [3].

As in this study we have only concern on the longitudinal dynamics, thus only the equations of motion that describe the longitudinal motion are presented. These set of equations of motion are with reference to the aircraft body-axes. Complete detailed descriptions of the aircraft can be further found in [3].

R. Akmeliawati and I.M.Y. Mareels: *Complex Stability Radius for Automatic Design of Gain-Scheduled Control Part II: Example on Autopilot Design*, Studies in Computational Intelligence (SCI) **76**, 251–258 (2007)
www.springerlink.com

$$\dot{x}(t) = f(x(t), u(t)), \tag{29.1}$$
$$y(t) = g(x(t), u(t)), \tag{29.2}$$

where $x \in \mathbb{D}^4 = [u_b \; w_b \; q \; \theta]^T$, the forward and downward velocity in the body-axes, the pitchrate [rad/s], and the pitch angle [rad], respectively; $u \in \mathbb{D}^2 = [\delta_e \; \delta_{th_1} \; \delta_{th_2}]^T$, the elevator and the left and right throttle deflection angle, respectively. $\mathbb{D} \subset \mathbb{R}$ specifies the region within the flight envelope of interest. $y \in \mathbb{D}^6 = [q \; n_z \; V \; \theta \; \alpha_b \; \gamma]^T$, and the dynamic equations, $f(t,x,u)$ are defined as follows.

$$f(t,x,u) = \begin{bmatrix} \{[(a_1 + a_2\alpha_b)\frac{w_b}{u_b} - (a_4 + (a_5 + a_6\alpha_b)(a_7 + a_8\alpha_b))] \\ p_8 u_b V_b + p_9 w_b q + g(\delta_{th_1} + \delta_{th_2} - \sin\theta)\} + d_6 w_b V_b \delta_e \\ \\ -p_1 w_b \alpha_b^2 - (p_3 w_b + p_2 u_b) V_b \alpha_b - p_4 u_b V_b \delta_e \\ -V_b(p_7 w_b + p_5 u_b) + (1 - p_6) q u_b + g\cos\theta \\ \\ -\{[(d_4 w_b + d_5 u_b)(a_1 + a_2\alpha_b) + (d_5 w_b - d_4 u_b) \\ (a_4 + (a_5 + a_6\alpha_b)(a_7 + a_8\alpha_b))]V_b \\ +[a_3(d_4 w_b + d_5 u_b) - d_9 V_b]q - (d_7\alpha_b + d_8)V_b^2 \\ +a_9(\delta_{th1} + \delta_{th2})\} + \{-V_b(d_1 w_b + d_2 u_b) + d_3 V_b{}^2\}\delta_e \\ \\ q \end{bmatrix}, \tag{29.3}$$

$\alpha_b = \arctan\frac{w_b}{u_b}$, $V = \sqrt{u_b{}^2 + w_b{}^2}$, $\gamma = \theta - \alpha_b$, $n_z = \dot{u}_b/g$ are the angle of attack, the airspeed, the flight path angle (FPA), and the vertical load, respectively. Constant a_n, p_n, and d_n, where $n = 1, 2, ..., 9$ are derived from the aerodynamics coefficients and the aircraft structure. The aircraft mass and the center of gravitation (COG) position are assumed constant.

The control objectives are to satisfy certain Flying Qualities requirement as in [4] and performance criteria summarized in Table 29.1.

The stability criteria follows the Flying Qualities requirement [4], whereas the performance criteria are summarized in Table 29.1. t_r, t_s are the rise time and the settling time, respectively. RMSE is the root mean square error of the response.

29.3 Controller Design

The structure of the automation program is shown in Fig. 29.1. Based on Fig. 29.1, the automation follows the following steps.

STEP 1: Choose scheduling variables. We select V and γ. Intuitively, V and γ are slowly time-varying and capture the dynamic nonlinearities, so they are prima facie candidates.

Table 29.1. Performance criteria

Responses	t_r(s)	t_s(s)	RMSE
Altitude	< 12	< 45	9.15 m
FPA	< 5	< 20	–
Airspeed	< 12	< 45	$< 1\%$
Pitch angle	< 12	< 45	$\pm 0.5\frac{\pi}{180}$ rad

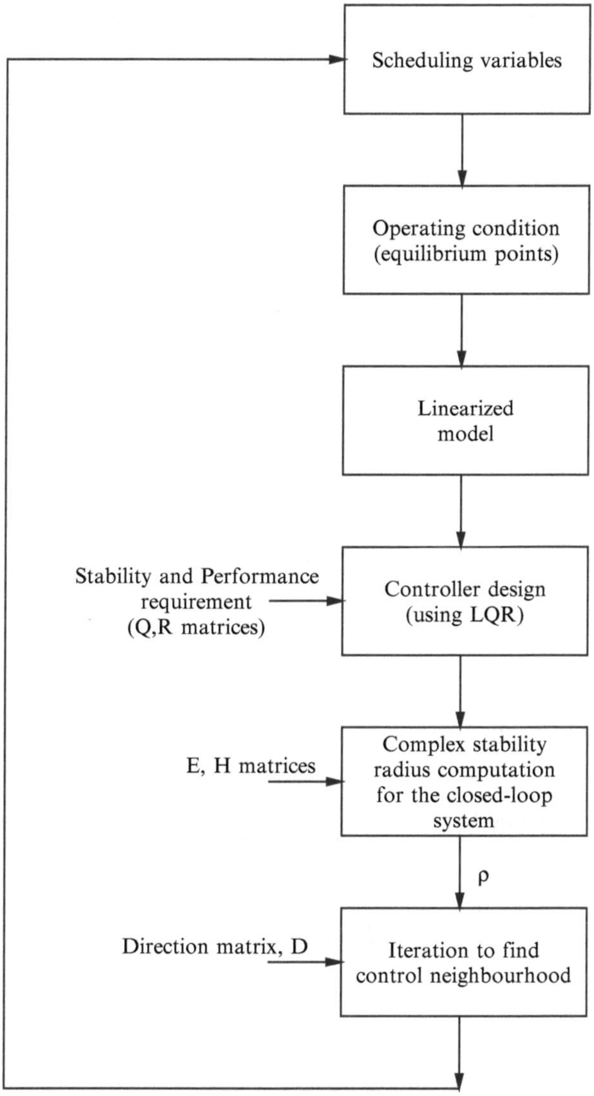

Fig. 29.1. The GS automation flowchart [1]

STEP 2: The first operating point is selected as $(\gamma_0, V_0) = (0, 80)$. This point is considered to be our nominal operating point.

STEP 3: Obtain the linear model. In this step, we linearize the aircraft model around the operating point selected in STEP 2.

STEP 4: Controller design. In this step, we design a feedback controller for the system defined in STEP 3, $u = -Kx$. For our nominal system, we use linear quadratic

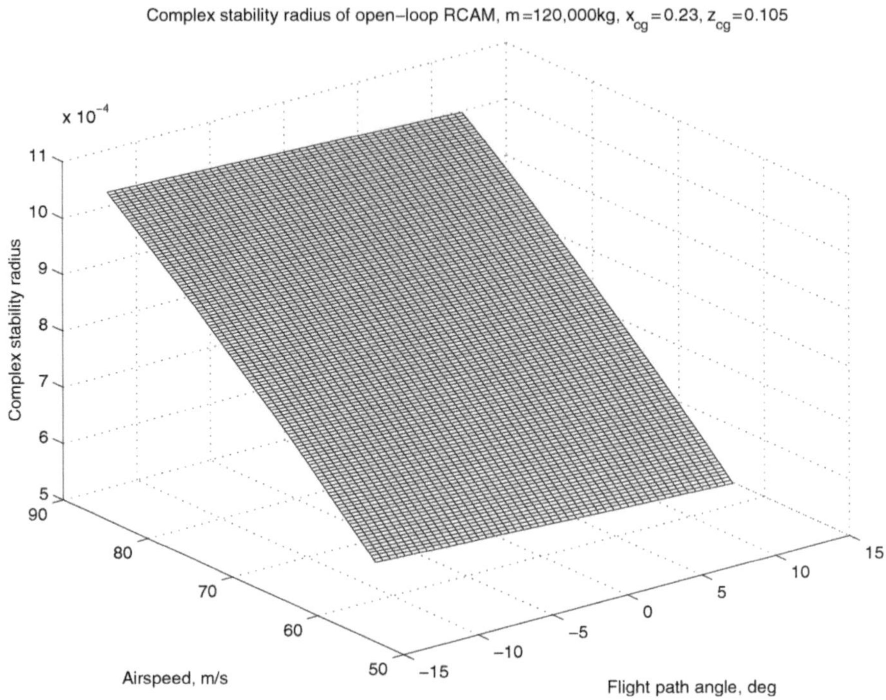

Fig. 29.2. Complex stability radius of open-loop system

regulator (LQR) method to design full-state feedback controllers. The weighting matrices are $Q = diag(0.01, 0.001, 0.001, 0.01)$, and $R = diag(10, 10, 10)$. The weighting matrices are chosen such that the stability and performance criteria can be achieved. The choice of Q is mainly aimed to improve the phugoid mode, which is characterized by u_b. The choice of R is to prevent unreasonably large control inputs.

STEP 5: Compute the (complex) stability radius. In this step, we first determine the structure of the perturbation matrices, E and H. The computed complex stability radius for the nominal operating point is 0.0374. The type of controller used affects the magnitude of the stability radius. Figures 29.2 and 29.3 show how the stability radii of the open-loop system over the entire operating region of interest have been enlarged significantly by adding the (LQR) controller[1] in STEP 5.

STEP 6: Iterate to find a neighborhood, i.e., the next operating points to be selected. The neighborhood of the point (0,80) in the form of a 'beehive' is shown in Fig. 29.4. This corresponds to eight directions, which are described by the direction matrix, D. The iterative procedure above is guaranteed to stop reasonably in finite time as $(A - BK)$ is Hurwitz.

[1] The same Q and R matrices are used for each operating point.

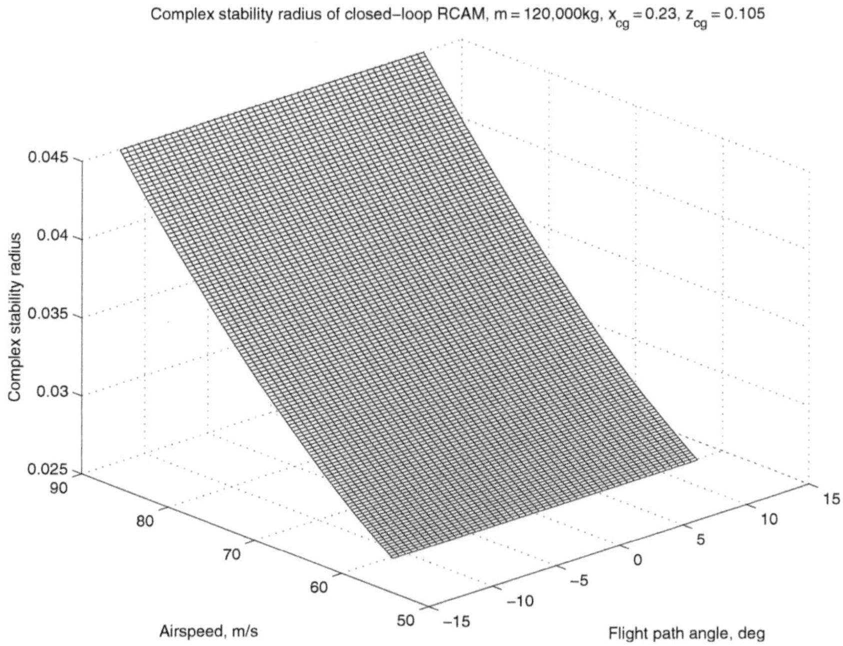

Fig. 29.3. Complex stability radius of closed-loop system

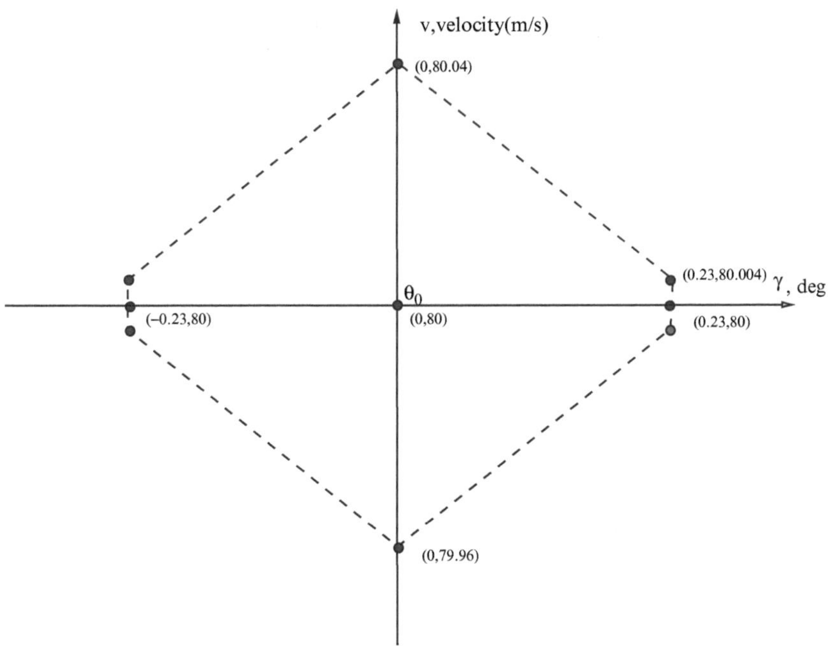

Fig. 29.4. 'Beehive' mapping of the stability region of point (0,80) in the scheduling-variable space

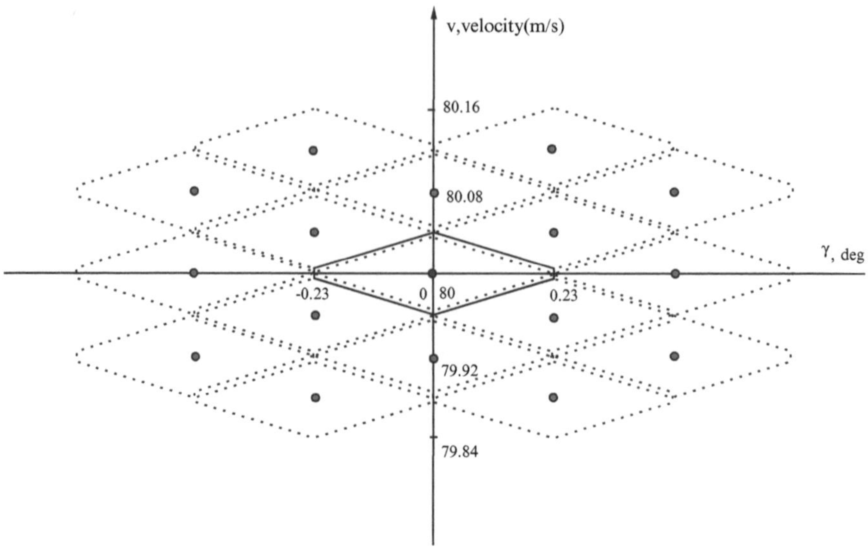

Fig. 29.5. The resulting grid

Figure 29.5 shows the resulting (γ, V) flight envelope grid. Our algorithm generates a partition on the basis of stability radius for the linearized system at each center. A formal verification requires a Lyapunov argument, which can be completed along the lines of Theorem 1 in Part I. Our result is conservative. This is due to the fact that the complex stability radius allows for a large class of time variations in the system, larger than the system robustness actually requires.

Notice also that the grid structure is quite regular. This regularity is not guaranteed a priori as the grid is the projection via a highly nonlinear mapping of a ball in parameter space into the scheduling-variable space.

29.4 Simulation Results

Using the resulting (γ, V) flight envelope grid (Fig. 29.5), we simulate the closed-loop systems to achieve trajectory tracking. Seventeen LQR controllers (based on 17 points of the grid) are computed using a MATLAB^{TM} program and are then interpolated using spline interpolation. The flight condition is normal, i.e., no wind disturbances. The desired trajectory is given. The initial operating point is $[u_{bo}, w_{bo}, q_0, \theta_0] = [80, 0.5, 0, 0.03]$. The output responses are shown in Fig. 29.6, whereas the control actions are shown in Fig. 29.7.

We observe that there is small tracking error in the state responses during climbing. However, the error is still within the design specifications. The maximum deviations in steady state of the airspeed and the pitch responses are less than 1% and 0.5°, respectively. The airspeed, the pitch, and the flight-path angle responses satisfy the stability and performance requirement.

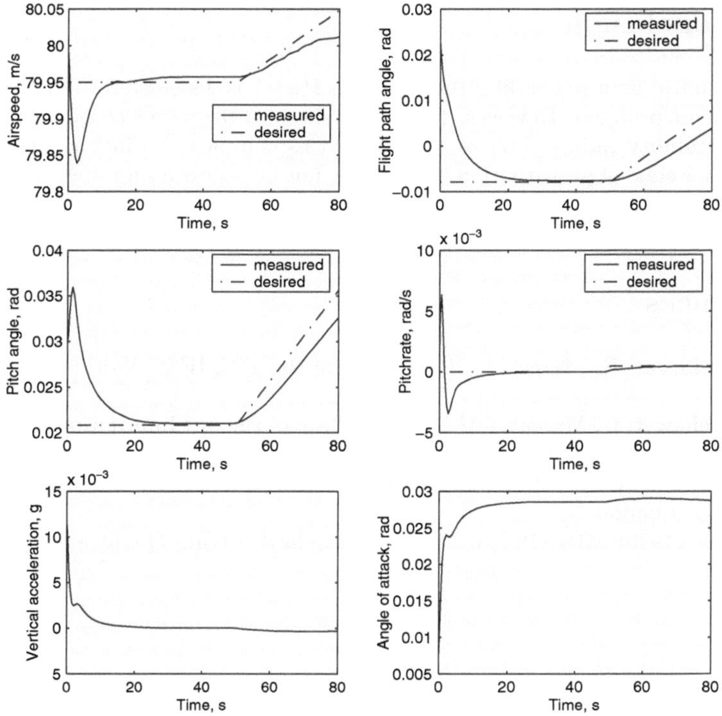

Fig. 29.6. Output responses

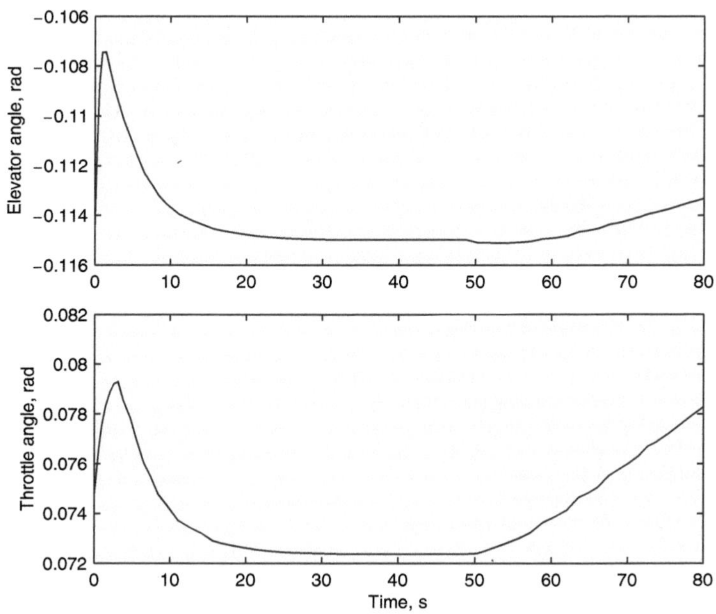

Fig. 29.7. Elevator and throttle activities

29.5 Conclusion

Our automatic gain-scheduling technique in Part I is successfully illustrated on a flight control problem. However, the result shows a degree of conservatism of the complex stability radius. This is as expected as complex stability radius can deal with a wider class of perturbation, which may not be necessary for our aircraft model in the example.

References

1. Akmeliawati R, Mareels I (1999) Proc. of the 14th IFAC World Congress 289–294
2. Akmeliawati R, Mareels I M Y (2006) Proc of the 3rd International Conference on Autonomous Robots and Agents (ICARA 2006) 665–670
3. Bennani S, et.al. (1997) Robust Flight Control, A Design Challenge. Springer-Verlag, London
4. USAF (1990) MIL-1797-A: Military Standard, Flying Qualities of Piloted Vehicles

30

Synthesis of Reconfigurable Hierarchical Finite State Machines

Valery Sklyarov and Iouliia Skliarova

Department of Electronics, Telecommunications and Informatics
IEETA University of Aveiro, Aveiro, Portugal
skl@det.ua.pt, iouliia@det.ua.pt

Summary. The paper suggests design methods for reconfigurable hierarchical finite state machines (RHFSM), which possess two following important features: (1) they enable the control algorithms to be divided in modules providing direct support for "divide and conquer" strategy; (2) they allow for static and dynamic reconfiguration. Run-time reconfiguration permits virtual control systems to be constructed, including systems that are more complex than capabilities of available hardware. It is shown that RHFSM can be synthesised from specification in form of hierarchical graph-schemes with the aid of the considered in the paper VHDL templates. The results of experiments show correctness of the proposed methods and their applicability for the design of engineering systems.

Keywords: Control systems, Hierarchical specification, Finite state machines, VHDL templates.

30.1 Introduction

Finite state machines (FSM) are probably the most widely used components in digital systems. This chapter suggests methods of FSM synthesis that possess two following characteristics important for robotics and embedded applications [1, 2]: (1) supporting the hierarchy and the strategy "divide and conquer" and (2) permitting statically and dynamically reconfigurable control systems to be designed.

The considered materials are derived from [6] where they were first presented and discussed.

30.2 Modular Specification of RHFSM

Modular specification can be done in the form of hierarchical graph-schemes (HGS) [3]. Figure 30.1 shows an example. Module Z_0 in Fig. 30.1 represents a top-level recursive algorithm for solving different optimisation problems over binary and ternary matrices [6].

V. Sklyarov and I. Skliarova: *Synthesis of Reconfigurable Hierarchical Finite State Machines*, Studies in Computational Intelligence (SCI) **76**, 259–265 (2007)
www.springerlink.com

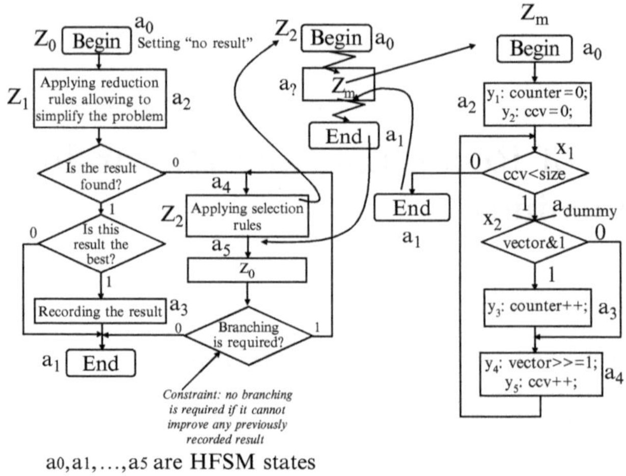

Fig. 30.1. Example of modular specification

One of the tasks of module Z_2 is executing operations over binary and ternary vectors. As an example, module Z_m in Fig. 30.1 describes a trivial algorithm for counting the number of ones in a given binary vector.

Note that for numerous optimisation tasks over matrices different algorithms have to be executed [6]. Thus, to provide adaptability of the module Z_0 for solving different problems it is necessary to reconfigure the module Z_m.

We will divide the considered task into static hardware implementation and dynamic reconfiguration of modules. Finally, we will show how to design hardware, which joins these subtasks and provides advanced operations for modules' collaboration.

30.3 Synthesis of HFSMs

Methods of HFSM synthesis from HGS specification were proposed in [3] and we will demonstrate these methods just on an example. The following VHDL code describes the stacks (called *M_stack* and *FSM_stack* [3]) that are used as HFSM memory:

```
process(clock,reset)
  begin                                                    (1)
    if reset = '1' then -- initialising
    elsif rising_edge(clock) then
      if inc = '1' then
        if -- test for possible errors
        else sp <= sp + 1; -- sp – stack pointer
            FSM_stack(sp+1) <= a0;
            FSM_stack(sp) <= NS; -- next state
            M_stack(sp+1) <= NM; -- next module
        end if;
      elsif dec = '1' then sp <= sp – 1;
```

```
      else  FSM_stack(sp) <= NS;
      end if;
    end if;
  end process;
```

Here, the signal *inc* is generated in any rectangular node of HGS, which calls another module; the signal *dec* is generated in the *End* node of any module.

The following VHDL fragment describes a template (skeletal code) for the combinational circuit of HFSM, and the module Z_m from Fig. 30.1 is completely specified.

```
process (CM,CS,X) -- current module (CM), state (CS)
  begin              -- X = {x1, ...,xL} - input signals
    case M_stack(sp) is
      when Z1 =>
        case FSM_stack(sp) is -- state transitions and
                       --output generation in the module Z1
        end case;
      -- .......................................................
      when Zm =>        -- below the complete code is given
        case FSM_stack(sp) is
          when a0 => Y <= (others => '0');
                     inc <= '0'; dec <= '0'; NS <= a2;
          when a1 => Y <= (others => '0');
                     inc <= '0'; NS <= a1;
                     if sp > 0 then dec <= '1';
                     else dec <= '0';
                     end if;
          when a2 => Y <= ''11000''; -- y1 and y2
                     dec <= '0'; inc <= '0';
                     if x1='0' then NS <= a1;
                     elsif x2='0' then NS <= a4;
                     else NS <= a3;
                     end if;
          when a3 => Y <= ''00100''; -- y3
                     dec <= '0'; inc <= '0';
                     NS <= a4;
          when a4 => Y <= ''00011''; -- y4 and y5
                     dec <= '0'; inc <= '0';
                     if x1='0' then NS <= a1;
                     elsif x2='0' then NS <= a4;
                     else NS <= a3;
                     end if;
          when others => null;
        end case;
      -- repeating for all modules, which might exist
    end case;
  end process;
```

The following VHDL code gives an example of hierarchical module call in the module Z_0:

```
when a2 => dec <= '0'; inc <= '1'; NM <= Z1;
```

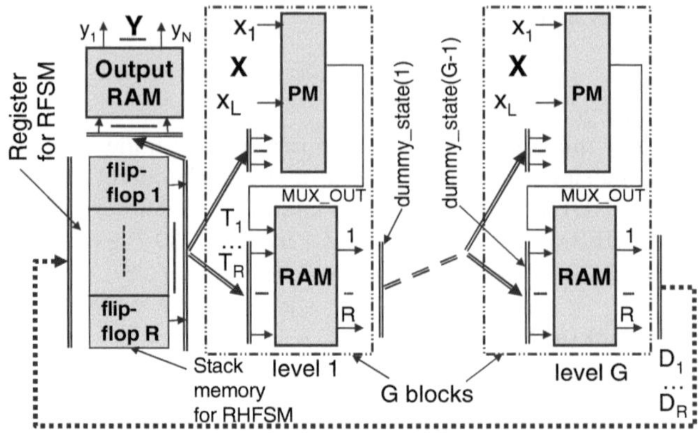

Fig. 30.2. Basic structure of RFSM

30.4 General Structure of Reconfigurable FSMs

Methods for synthesis of non-hierarchical reconfigurable FSMs (RFSM) were suggested in [4]. The cascaded RFSM model from [4] (see Fig. 30.2) is composed of RAM blocks, programmable multiplexers (PM) and a register.

By modifying the contents of the RAM blocks we can implement any desired behaviour within the scope of predefined constraints [4], which are the size R of the RFSM register, the number of RFSM inputs/outputs L/N and the number of reprogrammable levels G, which are the primary building blocks of the RFSM combinational circuit (CC). The PM is composed of a RAM and a multiplexer and is used for selecting appropriate input variables dependently on FSM states. VHDL code for the circuit in Fig. 30.2 was presented in [4].

30.5 Structure of RHFSM

Figure 30.3 depicts the basic structure of RHFSM, which permits to reconfigure the functionality of CC through reloading the RAM blocks. Changes to the modules can be provided through re-switching segments of the same RAM using the most significant bits (MSB) of RAM address space. Thus, MSB in Fig. 30.3 permits to select the current module (CM), and the less significant bits (LSB) of RAM address space are used in the same manner as in [4].

Note that RHFSM in Fig. 30.3 can be reconfigured just statically. To provide for dynamic reconfigurability it is necessary to supply additional functionality, which would allow reloading RAM blocks during execution time. Figure 30.4 presents a circuit, which implements such functionality. To reconfigure any module we need to reload 2G memory blocks. Indeed, to modify functionality of any level in Fig. 30.2 it is necessary to reload the RAM block for PM and the RAM block responsible for state transitions (this block is designated as RAM in Figs. 30.2, 30.3). Activating the proper RAM block is achieved through the respective enable signals (see Fig. 30.4). In order to simplify reloading, dual port RAM blocks have been used in such a way that the first port provides for normal functionality and the second port enables

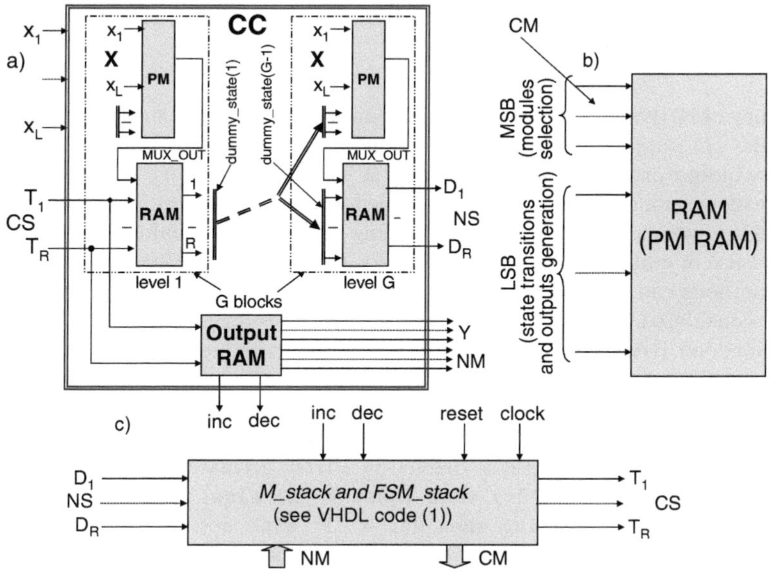

Fig. 30.3. CC of RHFSM (**a**), dual-port RAM block (**b**) and RHFSM stack memories (**c**)

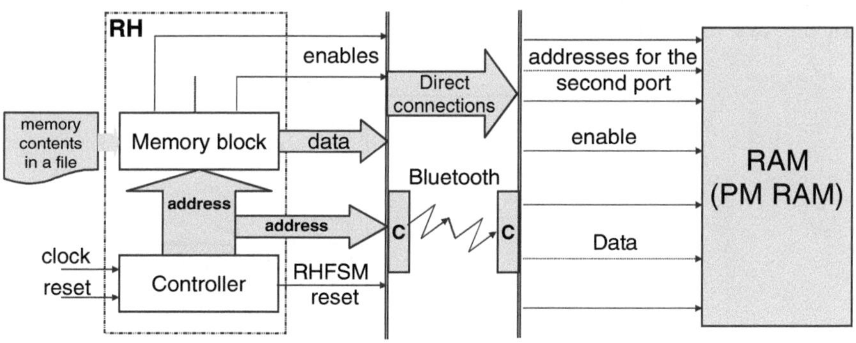

Fig. 30.4. Dynamic reconfiguration of RHFSM

the controller, as shown in Fig. 30.4, to change the contents of RAM. Two types of feasible reconfigurations (wired and wireless) have been considered. The first type, which is a wired reconfiguration, has been implemented and tested.

The reconfiguration handler (RH) in Fig. 30.4 is composed of a source for data and a controller. The controller generates memory addresses and copies data from the source to the RAM blocks activated by the appropriate enable signals. When all the blocks are programmed the controller resets the RHFSM and the latter is set into a working mode.

30.6 Advanced Techniques for Stack Operations in RHFSM

Efficiency of RHFSMs can be significantly improved through the use of the following methods: (1) supporting multiple entry points to sub-algorithms; (2) employing fast unwinding procedure for stacks used as an HFSM memory in case of recursive module invocations; (3) establishing flexible hierarchical returns based on alternative approaches, which can be chosen depending on the functionality required; (4) the rational use of embedded memory blocks for the design of RHFSM stacks. The first three methods can be used in the same manner as it was done in [5]. The last method will be considered in detail.

Embedded RAM can be efficiently employed to reduce hardware resources for the required stack memories. Figure 30.5 presents an example demonstrating how to describe process (1) in VHDL assuming that an embedded memory block (such as Xilinx library component RAMB4_S8_S8) is used for stack memory. Signal *sp_1* forms the RAM address and it is defined in VHDL architecture as follows:

```
sp_1 <= (others => '0') when reset = '1' else
        sp - unwinding when dec = '1' else sp;
```

As a result, the signal *sp_1* provides for fast stack unwinding [5] if required. Note that in Fig. 30.5 just a small part of the available memory block was used. That is why unnecessary address inputs were set to 0 and some outputs were left unconnected.

Since the stack pointer is common to both stacks, two segments of the RAM data bus are used for the *FSM_stack* and the *M_stack*, respectively. Two signals

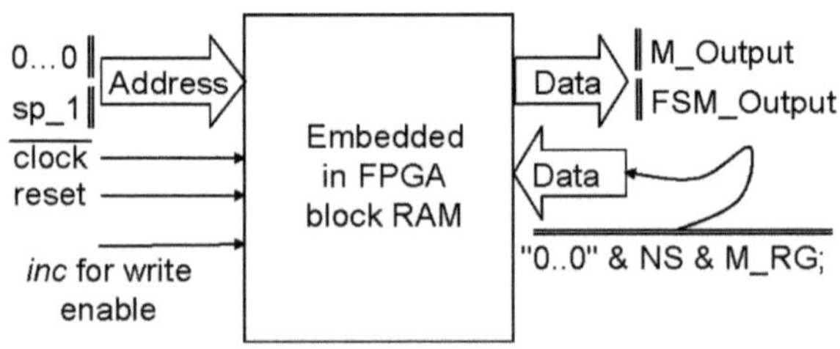


```
process (clk,reset)
begin
if reset = '1' then  -- initialization
elsif rising_edge(clock) then
    if inc = '1' then
        -- error handling
        sp <= sp + 1; FSM_RG <= NM_FS; M_RG <= NM;
    elsif dec = '1' then
        sp <= sp - unwinding; M_RG <= M_Output;
        FSM_RG <= FSM_Output;
    else  FSM_RG <= NS;
    end if;
end if;
end process;
```

Fig. 30.5. Embedded memory block

FSM_RG and *M_RG* enable the process shown in Fig. 30.5 to function at a single level. Switching to different levels of stack memories is provided through copying the RAM block outputs to *FSM_RG*/*M_RG*, which is needed just for hierarchical returns. This permits very fast HFSMs to be constructed.

The same approach can be used for distributed memory available for Xilinx FPGAs. The following code demonstrates how to use the distributed library component RAM16X1S:

```
RAM16X1S_instMO : RAM16X1S
    generic map (INIT => X"0000")
    port map (M_Output(0), sp_1(0), sp_1(1), sp_1(2),
              sp_1(3), M_RG(0),
              inverted_clk, -- inverted_clk <= not clock;
              inc);
```

N-bit stacks (N > 1) can be built from N components shown above.

Note that similar FPGA RAM blocks and distributed memory blocks can be used for all memories shown in Fig. 30.3. Thus, the proposed RHFSM is very well suited for implementation in commercially available FPGAs.

30.7 Conclusion

The chapter presents novel methods for the design of RHFSMs from hierarchical specifications. All the suggested structures are supported by VHDL templates (skeletal code, which can be used directly in engineering practice). The majority of the proposed methods have been verified in commercial hardware (in Xilinx FPGAs). The proposed RHFSMs have two very important features: (1) they support the design hierarchy (including possible recursive calls, if required) and (2) they are statically and dynamically reconfigurable, which is very important for adaptive systems.

References

1. Bojinov H, Casal A, Hogg T (2000) Emergent structures in modular self-reconfigurable robots. In: Proc. 2000 IEEE International Conference on Robotics and Automation, USA, pp 1734–1741
2. Meng Y (2005) A dynamic self-reconfigurable mobile robot navigation system. In: Proc. 2005 IEEE/ASME International Conference on Advanced Intelligent Mechatronics, USA, pp 1541–1546
3. Sklyarov V (1999) Hierarchical finite-state machines and their use for digital control. IEEE Transactions on VLSI Systems, 7(2):222–228
4. Sklyarov V (2002) Reconfigurable models of finite state machines and their implementation in FPGAs. Journal of Systems Architecture, 47:1043–1064
5. Sklyarov V, Skliarova I (2006) Recursive and iterative algorithms for N-ary search problems. In: Proc. 19th World Computer Congress, Professional Practice in Artificial Intelligence, Santiago, Chile, pp 81–90
6. Sklyarov V, Skliarova I (2006) Reconfigurable Hierarchical Finite State Machines. In: Proc. 3rd International Conference on Autonomous Robots and Agents - ICARA'2006, Palmerston North, New Zealand, pp 599–604

Author Index

MIX
Papier aus verantwortungsvollen Quellen
Paper from responsible sources
FSC® C105338

If you have any concerns about our products,
you can contact us on
ProductSafety@springernature.com

In case Publisher is established outside the EU,
the EU authorized representative is:
Springer Nature Customer Service Center GmbH
Europaplatz 3, 69115 Heidelberg, Germany

Printed by Libri Plureos GmbH
in Hamburg, Germany